A Naturalist's Guide to

Garden Birds

of the British Isles

A Naturalist's Guide to

Garden Birds

of the British Isles

PETER GOODFELLOW
PHOTOGRAPHIC CONSULTANT: PAUL STERRY

JOHN BEAUFOY PUBLISHING

First published in the United Kingdom in 2015 by John Beaufoy Publishing Ltd
11 Blenheim Court, 316 Woodstock Road, Oxford OX2 7NS, England
www.johnbeaufoy.com

Photo Credits:
Front cover: Blue Tit © Paul Sterry. **Back cover:** Robin © Paul Sterry. **Title page:** Barn Owl © Rob Read.
Contents page: Firecrest © Paul Sterry.

Main descriptions: photos are denoted by a page number followed where relevant by t (top), b (bottom), l (left)
or r (right).
Dave Ashton: 37b. **Frank Blackburn:** 66, 67t, 87l. **Laurie Campbell:** 29b. **Colin Carver:** 56tr. **Hugh Clark:**
64br. **Ernie Janes:** 13 (Sparrowhawk), 36. **Owen Newman:** 12 (Kestrel), 38, 89r. **Philip Newman:** 88l, 109t,
142. **W. S. Paton:** 29t. **Rob Read:** 3, 12 (Barn Owl), 42, 43b, 61r, 75t, 77b. **Richard Revels:** 53b. **Don Smith:**
63t. **E. K. Thompson:** 13 (Tawny Owl), 62. **Roger Tidman:** 12 (Grey Partridge), 27, 28, 108, 119b, 134, 145tl,
155b. **Steve Young:** 49tr, 153r.

ISBN 978-1-909612-35-8

Designed and typeset by Gulmohur Press

Printed and bound in Malaysia by Times Offset (M) Sdn. Bhd.

·CONTENTS·

INTRODUCTION

Many people today would not call themselves keen birdwatchers, twitchers or ornithologists. Nonetheless they have a keen interest in the countryside and the things that can be seen in their gardens – and perhaps when they are gardening or looking out of the windows of their homes, they see birds and wonder what they are.

The UK has a passion for feeding garden birds. The Royal Society for the Protection of Birds (RSPB) reports that over half of British adults feed garden birds. The British Trust for Ornithology (BTO) estimates that more than £200,000,000 is spent every year on bird food. Pet shops and garden centres throughout Britain stock their shelves with bird food, as well as feeding equipment such as birdfeeders and bird tables, and dozens of suppliers are listed on the Internet. Many farmers in recent years have started growing specialist bird seed for this trade. Despite this, many people do not know the names of the birds that come to their feeders. This book helps you get to know your garden birds better – you may find that there are far more species of birds in your garden than you realize.

The book is an introductory guide to help you identify the birds that you are most likely to see and hear, at any time of the year, in or above your garden. In all 90 species have been chosen, representing a wide variety of breeding species, and others from the far north and east especially that breed outside the area covered by this book, but visit Britain's shores on migration in spring and autumn.

USING THIS BOOK

Several ways to help find the right identification are provided. As well as a selection of photographs, there is information on each bird covering its description (see also p. 8), habitats, habits and distribution. Where possible, photographs that show the differences between the male and female, and summer and winter plumage, are included. The symbol (♀) printed on the photo denotes a female. Where there is no symbol, the photo is of a male bird.

These differences are further explained in the text within the descriptions. The information under habitat will help you to identify whether a particular habitat is near where you live. The comments on the birds' habits are selective: there may be an emphasis on particularly distinctive feeding habits (like those of the Spotted Flycatcher), or interesting information on breeding (as for the Rook). Each bird's voice is described – although the voice of a bird is notoriously difficult to put into 'words'. It may be useful for you to buy a disc with British birds' calls and songs (see, for example, the RSPB's catalogue), to help you to familiarize yourself with avian voices.

Under Distribution, the regions in Britain – the United Kingdom – where each species occurs, are given. The text also states whether the birds are resident, or summer visitors coming to Britain to breed but wintering elsewhere, or whether they nest outside Britain and are thus seen only on migration, generally arriving to spend the winter here.

BIRD CLASSIFICATION

The classification of birds has taxed the minds of scientists for centuries. Skeletal structure, plumage, voice and behaviour have all helped to put order into the 10,000 or so species in the world. Nowadays DNA testing is providing a new understanding of what a species is and its relationship to other birds.

For several hundred years birds have been named with words constructed from Latin and Greek. Order was given to the naming by the Swedish scientist Carl von Linné (Linnaeus is his Latinized name) with the publication of his book *Systema Naturae* in 1735. He brought order to nomenclature by deciding on a neat and tidy binomial system, instead of the many words that some scientists were using. Thus the House Sparrow became *Passer* (genus) *domesticus* (species, or the personal or specific name). Later it was found that an identifiable, geographic variation of the species required the addition of a third name; the Pied Wagtail of the British Isles, for example, became *Motacilla alba yarrellii* to distinguish it clearly from the continental bird, *Motacilla alba alba*, which we call the White Wagtail. Linnaeus had first named this species in 1735; the British zoologist Blyth gave the third name in 1834.

Unlike a common name, a scientific name belongs to one particular species across the world, no matter what language an observer may speak. Thus the scientific name *Passer domesticus* alludes to the same species whether its common name is House Sparrow (English), *Moineau domestique* (French), *Huismus* (Dutch), *Haussperling* (German), *Gråsparv* (Swedish) or *Домовый воробей* (Russian).

The order in which zoologists put birds is one that you will become more familiar with as you use the book more. Roughly speaking it goes from the most ancient, primitive birds (ducks and geese) to the most recently evolved ones (passerines, that is the songbirds). This orderly arrangement is constantly under review, so may appear slightly differently from one book to another. The taxonomic sequence used in this book follows that used by the BTO.

PLUMAGE AND STRUCTURE

Birds are identified by the colouration of different parts of the wings and body, as well as by shape. For example, just noticing that a bird has yellow wings is not enough, since the plumages of Yellowhammers, Goldfinches, Siskins and Greenfinches all contain some yellow.

The illustration on page 8 shows the names of the parts of a bird that are commonly used to describe its appearance. It is useful to become familiar with them. A longer, closer look at a bird on a bird table may show that its forehead is red with bands of white and black behind that, and that when it flies a broad yellow bar stretched across its mostly black secondaries and primaries can be seen – the only bird that fits this description is a Goldfinch.

eye-stripe
crown
supercilium
forehead
ear coverts
lore
nape
scapulars
mantle
chin
lesser coverts
throat
greater coverts
breast
back
flank
tertials
belly
vent
under-tail coverts
rump (base of tail)
tail
primaries
secondaries
carpal joint
axillaries
'fingers'

BIRDS MOST LIKELY TO BE SEEN IN GARDENS

There are dozens of families of birds, many of which are not found in Britain at all, such as vultures. Others are represented by only a few species compared with elsewhere – for example, Britain has only one kingfisher, but if you go to a holiday destination like The Gambia in West Africa, you will find that there are eight kingfisher species.

Seven bird families are well represented in this book: pigeons and doves, thrushes, warblers, tits, corvids (or crows), sparrows, finches and buntings. The birds you will see most often are in these families, apart from four other birds: the Dunnock, the only representative in Britain of the Accentor family, the Cuckoo, the only European species but with many others around the world, the Treecreeper, member of a small family, with one other European representative, another in North America and several in Asia, and the Starling, a member of the Sturnidae family with many more species elsewhere in both

Woodpigeon

House Sparrow

Blackbird

Great Tit

Chaffinch

Wren

Robin

mainland Europe and Asia. The BTO has been running a Garden Bird Feeding Survey since the winter of 1970–1971; it is the longest running survey of garden birds in Europe. Repeatedly, the most common birds in gardens are among the dozen species with the biggest populations in the UK. The British Trust for Ornithology has for several years been running surveys such as Bird Ringing, The Garden Birdwatch survey and the Breeding Bird survey, which give indications of our birds' populations. This data has been analysed by the organization's scientists and from it has come their latest estimate of the numbers of several species in the UK; it was published in BTO News, January/February 2013:

Wren	8,600,000
Robin	6,700,000
Chaffinch	6,200,000
Woodpigeon	5,400,000
House Sparrow	5,300,000
Blackbird	5,100,000
Blue Tit	3,600,000
Great Tit	2,600,000
Dunnock	2,500,000
Willow Warbler	2,400,000
Starling	1,900,000
Goldfinch	1,200,000

The top half-a-dozen birds seen in gardens vary slightly from year to year, but the same species are always high on the list. Get to know these birds well, so that you will know if you have a surprise newcomer. It is surprising how difficult it is for some people to observe carefully enough to identify a Blue Tit from a Great Tit, then *learn* the differences so that they do not forget which is which, and have to start all over again.

Birds that fly over gardens

Many garden birdwatchers are as excited to record a Swallow flying *over* their garden as seeing a bird *in* the garden. In this book birds such as the Swallow are included on the garden birds list, even though they do not nest in our garages or sheds (although some such birds may nest in certain gardens if the conditions are right). Some flyovers are regular: Swallows, House Martins, Swifts, gulls and corvids frequently appear over gardens. Every so often a surprise appears: it could be a flock of Lapwings, a bird of prey or a flock of Canada Geese honking their way from one river to another. Do record these observations and let your local bird club know about them (see p. 21 for more about this subject).

Red, Amber and Green Lists

For several years scientists have been looking for trends in our birds' populations to see how well they are doing. Many people – not just scientists – have been noting for quite a long time that some formerly common and widespread British birds, such as House Sparrows and Starlings, have become much more scarce and have disappeared altogether from certain areas. The RSPB, BTO and BirdLife International have created a list of British birds that have been carefully studied, whose populations have fallen, some by more than 90 per cent since the 1970s. The birds have been graded into three divisions. These are red (the grade including the most threatened birds), amber and green. There is no bending the rules. For example, the Nightingale population has declined by 49 per cent over the past 25 years, which puts it on the amber list. If it had declined by 50 per cent, it would be on the red list. In this book, red or amber list status is indicated in each species account where it applies. Several criteria are used to determine the status of each species:

RED LIST CRITERIA

- Globally threatened.
- Historical population decline in the UK during 1800–1995.
- Severe (at least 50 per cent) decline in the UK breeding population over the last 25 years, or a longer term period. The latter comprises the entire period used for assessments since the first Birds of Conservation Concern (BoCC) review, starting in 1969.
- Severe (at least 50 per cent) contraction of the UK breeding range over the last 25 years, or the longer term period.

Amber list criteria

- Species with unfavourable conservation status in Europe (SPEC, or Species of European Conservation Concern).
- Historical population decline during 1800–1995, but recovering; population size has more than doubled over the last 25 years.
- Moderate (25–49 per cent) decline in the UK breeding population over the last 25 years, or the longer term period.
- Moderate (25–49 per cent) contraction of the UK breeding range over the last 25 years, or the longer term period.
- Moderate (25–49 per cent) decline in the UK non-breeding population over the last 25 years, or the longer-term period.
- Rare breeder; 1–300 breeding pairs in the UK.
- Rare non-breeder; fewer than 900 individuals.
- Localized; at least 50 per cent of UK breeding or non-breeding population in ten or fewer sites, but not applied to rare breeders or non-breeders.
- Internationally important; at least 20 per cent of European breeding or non-breeding population in the UK (north-west European and East Atlantic Flyway populations used for non-breeding wildfowl and waders respectively).

GREEN LIST CRITERIA

Species that occur regularly in the UK but do not qualify under any of the above criteria.

A SELECTION OF SPECIES IN THE THREE CONSERVATION GROUPS

Red listed

LAPWING

GREY PARTRIDGE

MARSH TIT

SKYLARK

STARLING

SONG THRUSH

Amber listed

RED KITE

KESTREL

COMMON GULL

STOCK DOVE

BARN OWL

GREEN WOODPECKER

Green listed

MALLARD

SPARROWHAWK

MOORHEN

WOODPIGEON

HERRING GULL

TAWNY OWL

JACKDAW

MAGPIE

GOLDCREST

COAL TIT

BLACKCAP

WREN

MAKING A GARDEN ATTRACTIVE FOR BIRDS

It is really more appropriate to think of a garden as *wildlife friendly* in general, than to regard it as just a place to attract birds; this is the case whether the garden is tiny or covers several hundred square metres. Whether your garden is urban, suburban or rural it can be designed to attract birds. If it is close to farmland or woodland, or near the coast, there will be a bias in the types of bird that visit the garden towards the species that are in those habitats. Many birds are adaptable, and careful, quiet management of a garden could mean that if you live with it for 20 years or more, more than a hundred species may come your way. Some of these may visit every day, like Blue Tits, while others will occur perhaps just once, like maybe a calling Curlew flying over in autumn.

GARDEN PLANTS

We need to be aware not only of the birds in a garden, but also of the quality of the soil, the range and size of the bedding plants, the shrubs and trees the garden can accommodate, and all the insects and other invertebrates that the garden needs. Consideration must be given to how much of a garden is going to be devoted to hedges and flower borders, lawn (too much lawn turns a garden into a green oasis, especially if the grass is treated with weedkiller and insecticide), growing vegetables (Woodpigeons like to nibble the fresh leaves of vegetables), and fruits (Blackbirds love Blackberries and Raspberries as much as we do). Do the birds get a look in, or are they pests at harvest time? How much protection do the lettuces and soft fruits get?

When planning the planting, consider not only flowers and shrubs that provide seeds and berries as natural bird food, but also plants that are attractive to a variety and good number of insects. These small creatures form an essential part of the diets of many birds, especially as food for their chicks; and bear in mind that two of Britain's most charismatic birds, the Wren and Goldcrest, are insectivorous all year round.

Some people may think that their gardens are not big enough to be bird friendly – and there is currently a tendency to build high-density properties that include little room for gardens. However, imaginative thinking can, with the help of nest boxes (see p. 19), window boxes, bird baths (see p. 16), large tubs and hanging baskets, create mini gardens that do attract birds. Real surprises can happen: several times, Mallard ducks have been recorded nesting in window boxes, for example once outside the window of a fourth-floor apartment in Germany, and another time in Cardiff, Wales. Blackbirds will even nest in a creeper climbing around a regularly used kitchen door. The renowned economist Ernst Schumacher published a book in 1973 called *Small Is Beautiful*. His view of economics can be applied to your small garden, too, if you are thoughtful and determined to make the space a success.

If you have the space to attract a good range of insects to your garden, some of the plants you would do well to grow for their flowers are marigolds, Bell Heather, asters such as the Michaelmas Daisy, Ox-eye Daisy, ivies, honeysuckles, foxgloves and snapdragons. If you have room for a tree or two, native types are best such as Elder, hawthorns, rowans, Silver Birch and oaks, all of which produce flowers that attract insects like bees and hoverflies,

Country Garden

Goldfinch on Teasel

and fruits that are excellent food for birds. They do grow tall, so be prepared to do some heavy pruning after ten years or so. Three long-established non-native shrubs are excellent, too: Common Lilac (a member of the olive family from the Balkans), Californian Lilac, and Buddleia from America or Asia, the so-called 'butterfly bush' because butterflies are greatly attracted to its flowers.

Think about providing a wild area in your garden where you can grow plants such as Common Nettle, which attracts numerous insects. The insects in spring and the plants' seeds in autumn are both of great value for birds. Teasels produce spiky seed-heads that are of particular interest to Goldfinches – the very sharp bill of this species has evolved to extract seeds from the heads.

PROVIDING WATER

Some people say that the most important part of a bird-friendly garden is clean water. Birds need water to drink and to bathe in throughout the year. A simple bird bath will do, or if you have room, a pond; if you can arrange it, the sound of running water is particularly attractive to many species. A tray containing water will be sufficient for many birds; more than that depends on how much time and money you are prepared to spend. Note that bird baths should not be too deep, and should allow small birds to use them. Birds do not submerge their entire bodies, but only dip their wings and splash water onto their backs. Arrange to have gently sloping sides to both bird baths and ponds to enable birds to enter the water. A rough surface will prevent birds from slipping. Bear in mind that the water in a bird bath should be changed each day to avoid algae and bacteria that cause disease from forming in it.

PROVIDING SUPPLEMENTARY FOOD

As already noted, a huge amount of bird food is available to buy. Supplementary food in gardens is particularly useful in winter, when natural food sources may be in short supply, as well as in the nesting season, when birds are feeding their young. During late autumn and

Niger Seeds

Black Sunflower Seeds

Peanuts

Mixed Birdseed

winter high-fat foods such as peanuts, cheese and fat balls are helpful. In spring and summer high-protein foods such as mealworms help birds with young chicks.

Foods like seeds and grain can be bought in bulk and stored in lidded containers, which should be kept in a cool, dry place. Examples of seeds and grains available are black sunflower seeds, sunflower hearts, niger seeds (favoured by slender-billed birds such as Goldfinches, Siskins and redpolls), hemp seeds, millet seeds, wheat, barley and rolled oats. Peanuts are a good food for birds, but note that they can choke young chicks, so always place them in wire-mesh feeders that will prevent birds from taking away whole peanuts. Peanut butter, perhaps with nuts mixed in, can be put in a log that has had holes drilled into it.

Household scraps cut up into tiny pieces can be valuable food for birds. They include cheese, brown bread (white bread has little nutritional value), cooked rice and pasta (not

Common garden birds at feeder

cooked with salt, which can kill birds) and cat food. Clear away any uneaten food regularly to prevent it from spoiling. However if you are near the coast, scraps thrown on the bird table or lawn may lie only minutes before the Herring Gulls or Black-headed Gulls find them! You can place it in a mesh-wire hanging basket to contain it. Note that windfall fruits are very valuable to birds such as thrushes during cold-weather spells.

A variety of birdfeeders and bird tables is available. Squirrel-proof birdfeeders with a metal cage around the feeder are useful in areas where squirrels are an issue. Cage birdfeeders allowing access only to small birds, but excluding larger birds such as pigeons, are available. Both birdfeeders and bird tables should be positioned near high cover to provide concealment from birds such as Sparrowhawks. All birdfeeders should be taken apart and cleaned regularly, as should bird tables. Wear gloves when cleaning them (and bird baths).

NESTS IN GARDENS

One of the joys of birds in the garden is having a pair or more of birds. Opportunities for breeding can be provided by growing suitable trees and shrubs that provide safe cover and support for a nest, or by erecting artificial sites in the form of nest boxes.

Hawthorns, privet, cotoneasters, small conifers and climbing plants are all good for Blackbirds, Song Thrushes, Dunnocks, Wrens and finches. It is amazing how secretive birds are when building and rearing a brood. Many people (birdwatchers included) have not discovered a nest in a bush or hedge until the plant is being pruned later in the year. This does imply that work on a bush or hedge should not be done until you have looked

Common style wooden bird box   *Typical open-fronted bird box*

in it carefully, and that in fact it is best if it is carried out at a time when the birds are not nesting – note that several species have broods late into the summer.

Many species readily use nest boxes, and different species need different types of nest box. A variety of nest boxes is available to buy. The most common type is an enclosed box with an entrance hole with a diameter made to suit the bird. Enclosed boxes are used by birds such as tits, Nuthatches, Starlings and Spotted Flycatchers. Robins and Pied Wagtails use a box with the top half of the front cut away. Spotted Flycatchers and thrushes use open-fronted boxes with low fronts. You can tempt House Martins and Swifts to colonize your house by erecting artificial nests. For House Sparrows try erecting a terrace of boxes. Make sure you buy or build a box which has a lid which you can lift (and later shut securely), so that the box can be cleaned at the end of the season.

Kestrels may nest in a large, open-fronted box fixed on a pole, tree or building at least 5m from the ground. They need quiet and undisturbed sites close to grassland for hunting, so will only nest in large, quiet gardens. Tawny Owls may nest in Kestrel boxes, or they may use a long, thin box fixed to a tree at an angle or under a sloping branch, with nearby branches for the chicks to use after they have hatched. The owls are sensitive to disturbance and can be aggressive during nesting, so gardens with pets and children are generally inappropriate. Barn Owls adapt readily to nest boxes, and may also use them year round for roosting. A Barn Owl box should have a visible hole and a ledge underneath to prevent the fledglings from falling out. It should be sited in a quiet place at least 1km away from major roads (the species hunts on roadside verges, so is vulnerable to traffic), with a clear view of hunting habitat such as open fields. Barn Owl boxes are thus best suited to large rural gardens.

Blue Tit at nest box

Erect a box so that the birds have a clear flight path to it, and so that it is sheltered from bad weather and ideally not facing strong sunlight – this means that in Britain a nest box should face somewhere from north to south-east. Make sure that a nest box is watertight from above, and that it has easy access from the top for cleaning – but that the lid cannot be lifted by a predator such as a cat. A metal plate around the hole will deter squirrels from entering and attacking the nestlings. Avoid boxes with perches, which may provide access to predators. If a box has been used it should be cleaned out in the autumn to ensure that no nest parasites are left to attack the following year's brood. Ideally, wait until at least October before cleaning out boxes to ensure that multi-brooded birds such as sparrows have finished raising their offspring (note that it is in fact illegal to disturb nesting birds). Wear gloves when cleaning nest boxes. (See p. 157 for source of supply.)

Bird conservation

It has already been noted that many species in Britain are rarer than they used to be. Our gardens have become one of the country's most important habitats, so we need to help the birds as much as possible with bird-friendly gardening – trying not to use insecticides or slug pellets, planting suitably and feeding responsibly.

Other ways in which birds can be helped include:
- Keeping notes of what birds come to the garden and when.
- Joining a local RSPB group and/or county bird society.
- Sending in records for the society's annual report.
- Joining the BTO Garden Birdwatch survey.
- Taking part in the annual RSPB count, which takes just an hour on a date in January.

Thousands of people feed birds, so are a potential band of researchers that can help provide data concerning the populations and breeding success of our birds. The BTO Garden Birdwatch survey and the RSPB count are ideal vehicles for turning your pleasure into science that will support conservation, and ultimately enable conservation organizations to successfully lobby parliament in order to improve Britain's protection laws, and persuade local authorities to obey these laws.

The BTO and RSPB may be found at their home websites – see p. 156, or write to:

BTO, The Nunnery, Thetford, Norfolk IP24 2PU.
RSPB, The Lodge, Potton Road, Sandy, Bedfordshire SG19 2DL.

Addresses for county or district ornithological societies can be found on the internet or from your local library.

More than 16,000 gardens are now in the BTO survey, which had its first full year in 1995. The participants regularly fill in, at the same times each week, coded forms that are read automatically. Record sheets cover a 13-week period, and can be posted or submitted over the Internet. Scientists at the BTO analyse the data and publish regular reports for participants. The survey has helped to show, for example, the real increase in the numbers of Goldfinches and Woodpigeons, the decline in House Sparrows, and that there are differences in the status of several species in different parts of the UK. The data also clearly show the rise and fall of the number of gardens visited by regulars like Chaffinches in winter and summer, compared with other years. Facts like these are vital for an understanding of the state of Britain's birdlife.

Sometimes you may notice that one of the tits or finches that come to your garden has a shiny metal ring, and perhaps a coloured ring, too, on its leg. About three-quarters of a million birds are caught and ringed each year in the UK by specially trained and licensed birdwatchers. Each metal ring has a unique number. If a ring found on a bird can be read and reported with the date and place of discovery to the BTO by post, phone or over the Internet, the particular bird's age will be revealed, which will help in the assessment of the species' survival rate and movements. If you regularly see a ringed Blue Tit, for example, someone near you is almost certainly the ringer, because Blue Tits do not travel far. Your local bird club may help you track down the ringer. If you are able to report a number on a dead bird's ring, the BTO will send you details of the bird's ringing history.

GLOSSARY

Arm The inner wing, the part of the wing nearer the body, inside the bend of the wing, the secondaries and their coverts.

Arboreal In regards to birds, living in trees.

Bare parts A bird's beak, legs and feet.

Broadleaved Trees with broad leaves, i.e. trees other than conifers, such as oak, ash, birch.

Carpel The bend in the wing, often called the wing-bend or wrist.

Cere A fleshy or waxlike swelling at the base of the upper part of the beak on certain birds, e.g. pigeons, birds of prey.

Colony A group of several pairs of the same species nesting close together; so, colonial.

Conifers Typically evergreen trees, with needle-like leaves and bearing cones as fruit, such as pine, spruce, cypress, yew and fir.

Crepuscular Of birds, active at twilight or before sunrise.

Cryptic Colouration or markings that serve to camouflage a bird in its natural environment.

Deciduous Of trees, shedding or losing leaves at the end of the growing season, in autumn.

Defoliating Depriving a plant or tree of its leaves.

Diurnal Active during the daytime.

Eclipse Of male members of the duck family, cryptic plumage attained in late summer, while moulting their flight feathers and becoming flightless.

Endemic Of birds and other creatures, native to or confined to, a certain region, so having a comparatively restricted distribution.

Falcon A bird of prey with long, pointed, powerful wings, capable of swift flight, e.g. Kestrel, Peregrine; term used by falconers for the female of the species (see also *Tiercel*).

Fledgling A young bird that has acquired its flight feathers and has just left the nest.

Flight feathers The strong feathers which form the outer part of the wing on the hand; the pinions or primaries, secondaries, tertials.

Fore-wing The upperwing coverts on the arm of the wing.

Frugivorous Fruit eating.

Genus (plural **genera**) In the scientific ordering of creatures it is the category below family but above species; in the international Latin name of the creature it is the first word.

Gregarious Tending to move in or form a group of the same or similar kind, so a flock.

Hand The part of the wing outside the wing-bend, especially the primaries.

Immature A bird wearing any plumage other than the adult.

Insectivorous Eating insects.

Juvenile A young, fledged bird wearing its first set of true feathers.

Local Living in a certain locality, not widespread.

Mask A dark colouring of the bird's head, from the beak, around the eyes.

Mirrors The white spots in the black near the tips of the longest primaries of a gull's wing.

Morph A variant of a specified colour within a species that is not geographically defined (as a subspecies would be). The 'phase' is used to mean the same, as in grey phase and brown phase, e.g. of a young Cuckoo.

Nocturnal Active at night, as opposed to diurnal.

Omnivorous Eating all kinds of food, vegetable and meat.

Orbital ring Circular area of bare, unfeathered skin surrounding an eye.

Phase See *Morph.*

Race A population of a species that differs from others of the same species in one or more characteristics, e.g. Pied and White Wagtails.

Raptor A bird of prey, particularly applied to eagles, falcons and hawks.

Resident Living in a particular region, not migrating, travelling only to find food or to roost.

Species A group of organisms capable of interbreeding and producing fertile young.

Speculum A bright, glossy patch of colour in the secondary feathers of ducks.

Subspecies A subdivision of a species often characterized by a variation in size and/or the colour of the plumage, e.g. Pied and White Wagtails; they do interbreed.

Subterminal Coming nearly at the end.

Terminal At the end.

Territory An area inhabited by a mated pair that is often vigorously defended.

Tiercel The male bird of prey, of which the female is the falcon.

Trailing edge The rear edge of the wing feathers.

Vagrant A species occurring well outside its normal range, often as a result of exceptional weather.

Wing-bar Contrastingly coloured, often much lighter, bar on the wing, often formed by light tips to the wing-coverts.

Wing-panel A contrastingly coloured area on the wing, which is but imprecisely described as far as size is concerned.

Wing-span (WS) the distance from wing tip to wing tip.

MALLARD ■ *Anas platyrhynchos* Length 50–60cm; wingspan 81–95cm
Amber listed

DESCRIPTION This is one of the 'dabbling ducks' that up-end for food – head under water, tail in the air. The male has a very distinctive pattern that includes a shiny green head, yellow bill, and purple, grey and black body. Its unique central tail feathers curl distinctively upwards, and it has a white neck-ring and orange legs. The female is camouflaged brown with darker markings, and has an orange bill marked with black, and a striking dark blue speculum with black-and-white borders. The species moults after breeding, when the drake resembles the female but has a yellowish bill. This is the ancestor of the 'farmyard duck', with many variations of colour, from pure white to shiny black.

HABITAT Occurs on lakes, ponds, canals and streams, from moorland to the coast.

HABITS The Mallard is very adaptable, nesting in city parks (it has even been known to nest in the window box of an apartment and in a flower pot in the beer garden of a Birmingham pub!). The species is omnivorous. The female lays 9–12 greenish-buff eggs in a nest constructed in thick vegetation and lined with the duck's down. The drake takes no part in rearing the brood. Voice: the duck quacks, while the drake has a rasping call and a low whistle.

DISTRIBUTION One of Britain's most widespread species, the Mallard is common throughout the region from the Orkneys to the Scilly Isles, in summer and winter; where there is water there will usually will be Mallard. The species is largely resident, but retreats from ice. It is most common from north-west England to the Thames Basin. British-bred birds are largely sedentary; in winter numbers are swollen by immigrants from continental Europe.

duckling

♀

CANADA GOOSE ■ *Branta canadensis* Length 90–100cm; wingspan 160–175cm

DESCRIPTION This is a large brown goose with a black head and neck, relieved by a contrasting white blaze from behind the eye that meets under the chin. The wings are dark above and below, the tail is white under the base, and the bill and legs are black.

HABITAT In the breeding season occurs on lowland lakes, even in city parks; found on large lakes and estuaries in winter.

HABITS Feeds on aquatic and waterside vegetation. The female lays 5–6 white to creamy-white eggs in a shallow depression lined with plant material and down. The gander defends his breeding territory aggressively, but the species is gregarious outside the breeding season, when flocks of hundreds or more birds form. The birds signal their approach in flight with deep, loud, trumpeting calls. This is when it is possible to see them over gardens; where a garden borders a river or lake, more birds may be seen. Canada Geese may become pests in large numbers, fouling lakesides and riverside lawns.

DISTRIBUTION Native to North America, and introduced by landed gentry to British estates in the 17th century. The species is now widespread due to lack of control since the mid-1900s, throughout England and Wales, where there has been a spectacular increase since the 1970s. It is patchier in Scotland, and there are few birds north of a line from the Isle of Mull to Perth. In Northern Ireland it is mostly confined to the valley of the River Erne. It has evolved to become rather sedentary, in contrast to North American long-distance migrants; its distribution in Britain is thus similar all year round.

RED-LEGGED PARTRIDGE ■ *Alectoris rufa* Length 32–34cm; wingspan 47–50cm

DESCRIPTION The sexes are similar in this species, with brown upperparts, a warm buff belly, and a lavender-grey crown, breast and flanks, the last vertically barred with white, black and chestnut. The supercilium is white, and a black eye-stripe forms a border around the white throat, the black breaking down into a bib of streaks. The bill and legs are red, and the red outer-tail feathers show in flight.

HABITAT Occurs on large, modern, open fields, and sand and gravel works, seldom wandering more than a few kilometres from where it was bred.

HABITS The Red-legged Partridge is often seen in coveys that are sometimes as large as several dozen birds. When disturbed the birds often run away, rather than flying away. The male regularly perches on a vantage point in his territory, such as a straw bale or fence post. The diet consists of seeds, leaves and roots of farmland plants; the chicks especially eat insects. The male makes a scrape in the ground, in which the female lays 10–16 eggs. Two broods are often reared, and the male often incubates the second clutch while the female looks after the first one. Both parents tend the young, calling quietly and constantly to keep the family together. Voice: a loud, rather hoarse, rapid triple *che-ke-ke, che-ke-ke, che-ke-ke*, with emphasis on the second and third syllables each time.

DISTRIBUTION A resident species, the Red-legged Partridge was introduced to Britain as a game bird in the 17th century, although the introduction was not successful until 1790 in Suffolk; it is native to France and Iberia. Since its introduction it has spread over much of England and eastern Scotland. It is widely found where there are also Grey Partridges, but it is not thought that the decline of that species has anything to do with the Red-legged Partridge. Over six million birds are reared and released each year on the many estates that have annual shoots.

GREY PARTRIDGE ■ *Perdix perdix* Length 29–31cm; wingspan 45–48cm

Red listed

DESCRIPTION A distinctive 'round' game bird, the Grey Partridge has an orange-brown head, with more orange on the male than on the female, a grey neck and breast, and chestnut-barred flanks. The upperparts are brown, the wings are marked with chestnut-brown and there is a dark brown horseshoe-shaped mark on the breast, which is less obvious on the female than on the male. The outer tail is rufous, and the bill and legs are blue-grey.

HABITAT Occurs principally on low-lying agricultural land with rough cover; also heaths and sand-dunes nearby.

HABITS When flushed, Grey Partridges fly low over a short distance on whirring wings. Their food consists chiefly of plant material, and chicks are fed on insects. They nest on the ground in good cover, and lay 10–20 olive-brown eggs. Both adults tend the chicks, which run easily from the day of hatching. The birds remain as a family until August, and occur in small flocks (coveys) from then until around February. Voice: the call sounds like a rusty gate.

DISTRIBUTION A mainly resident bird, the Grey Partridge is native to Britain (and widely in mainland Europe). It is widespread in England and the eastern lowlands of Scotland, and absent from most of Wales, Northern Ireland and south-west England. It is thinly spread in most of its range, and most abundant in Norfolk, Lincolnshire, Lancashire and eastern Scotland. More than 100,000 birds are reared and released every year on estates, which earn money from organized shoots. Despite this, as a wild bird this is one of the most strongly decreasing birds in Europe – it has experienced a 91 per cent decline since the late 1960s. This is due to intensification of farming practices, especially the use of herbicides on the food plants of the invertebrates on which the chicks feed.

COMMON PHEASANT ■ *Phasianus colchicus* Length 53–89cm; wingspan 70–90cm

DESCRIPTION The male is larger than the female, and has a dark, metallic green head with red wattles; some individuals have a white neck-ring. The body is usually chestnut-brown, covered with black chevrons, and the tail is over 35cm long and barred black. The female is buffish-brown covered with dark marks, especially on the upperparts.
HABITAT Resident on farmland, parkland and plantations. This is a commonly reared game bird; near estates that run shoots it can be seen in gardens.
HABITS The species is omnivorous, but feeds particularly on grain and other seeds, wild fruits and insects in the summer. The breeding male often has a harem of two or three

hens, and has little or nothing to do with incubation or chick rearing. The hen lays 8–15 olive-brown eggs in a scrape in the ground. Pheasants are wary and run for cover rather than fly, but they do become tame if they are fed regularly in gardens. They occur in flocks in winter. Voice: the breeding males crow loudly in spring; this call is often followed by brief, loud wing-flapping. They utter a harsh *kuttUk-kuttUK-kuttUK* when flushed.
DISTRIBUTION Common and widespread. Pheasants were introduced into England in the 11th century by the Normans, or perhaps even earlier by the Romans, and into Scotland and Wales in the 16th century. The introduced birds were native to the Caucasus of Russia; those with a white neck-ring are descendants of birds from the Far East, which were introduced later. They are found across Britain except in north-west Scotland, the Western Isles and Shetland. The most densely populated areas are in England, and eastern Scotland as far north as the Moray Firth. Pheasants are widely reared on estates for the autumn shoot; about 15 million are shot annually.

♀

GREY HERON ▪ *Ardea cinerea* Length 84–102cm; wingspan 155–175cm

DESCRIPTION The adult Grey Heron has very long legs and a long neck. The back and coverts are blue-grey, the flight feathers dull black. The head is white with a black stripe over and behind the eye, ending in long nape plumes, while the body is greyish-white with black-and-white stripes down the centre; note the adult's head pattern. The bill is dagger-like and yellow. Juveniles are much greyer than adults, and lack the black-and-white contrast on the head; the bill in juveniles is greyish-horn.

HABITAT Occurs on the shallow edges of lake sides, slow-flowing rivers, marshes and estuaries.

HABITS The flight of a heron is very distinctive: the neck is retracted in flight, and the legs trail beyond the tail; the wingbeats are deep and slow. Herons feed mostly on fish (including eels), also preying on amphibians, small birds and mammals. There are many records of Herons raiding garden ponds in suburbia for the goldfish. Be aware! They are usually solitary, but breed colonially in tall trees, in heronries. They lay 3–5 greenish-blue eggs in a bulky stick nest. Voice: this is often heard, the call being a loud, harsh *frank*.

DISTRIBUTION Grey Herons are widespread across the UK except in most mountainous areas. The BTO heronry census has been running since 1928, and is the longest running monitoring set of data for any bird species in the world. It shows that in the early 2000s there was a bigger population of Grey Herons than at any other time in the census. Harsh winters can cause severe mortality in the birds due to the freezing of wetlands – the source of their food.

juvenile

COMMON BUZZARD ■ *Buteo buteo* Length 51–57cm;
wingspan 113–128cm

DESCRIPTION The upperparts of Common Buzzards are dark brown, and the underparts are white with very variable brown streaks and bars, most noticeable across the breast. The underwings are pale with a dark carpal patch. Be aware that an eagle is half as big again and in the UK is confined to Scotland.

HABITAT These birds are most often seen spiralling over their territories on farmland, and in forest clearings, moorland edges with mature trees and even well-wooded suburbia.

HABITS Common Buzzards glide with their wings in a noticeable 'V' shape. They often hunt by waiting for prey passing below their perches. Their diet consists mostly of small mammals, especially Rabbits, and they also feed on carrion. The female lays 2–4 white eggs with variable red or brown markings in a bulky tree nest of sticks with a soft lining. The nest is often used for several years running, or a pair may use 3–4 nests in rotation. Voice: heard throughout the year, a mewing, far-carrying *pee-oo, pee-oo*.

DISTRIBUTION Most Common Buzzards are resident in Britain. By the 1900s persecution since the 18th century, pesticides and disease had greatly reduced the population, and the species was confined to Scotland, Wales and south-west England. Since then there has been an amazing change and Common Buzzards are now found throughout Britain. They are still most common in Wales, southern and south-west England, and southern and eastern Scotland. An increase in Rabbits, the withdrawl of harmful pesticides and a reduction in persecution have been key factors in the species' recovery. It rivals the Kestrel as Britain's most common bird of prey.

RED KITE *Milvus milvus* Length 61–72cm; wingspan 140–165cm

Amber listed

DESCRIPTION This is a large and distinctive bird of prey; it is noticeably long winged and has a characteristically long tail that is forked, most obviously when it is spread. The greyish-white head contrasts with the deep rufous body and rufous-orange tail. The underwing pattern is very distinctive, with the black wing-tip bordered by a large white 'window' on the inner primaries. The upper wing is rufous-brown, often with a buff band across the greater coverts.

HABITAT Small to medium woods, in rolling, arable and grassy farmland.

HABITS The Red Kite is a graceful, buoyant flier, constantly twisting its tail as it glides. Its diet consists mainly of carrion, as well as refuse, small mammals, birds, insects and other invertebrates. Nest building starts in a tree in early spring, and the female lays 1–3 white eggs with variable amounts of reddish-brown spots. Red Kites are often seen singly, but do gather in large numbers where there is plenty of food, especially at feeding stations on reserves and in private gardens. Voice: utters a shrill, mewing *peee-ow*.

DISTRIBUTION This species is a major conservation success. Kites were intensely persecuted as vermin from the 17th century onwards (Shakespeare's tragic hero King Lear speaks of the 'detested kite'), last bred in England in the late 18th century and were almost extinct in Britain after the Second World War; a small remnant, protected population survived in central Wales. A strong conservation lobby resulted in the careful reintroduction of Red Kites, with Swedish birds being used in 1989, and others later from Spain and Sweden, the last in 2010. These birds first bred successfully in 1992. Red Kites are now found widely from Wiltshire to Scotland, with particular clusters occurring around the introduction sites. Successful introductions include one in Northern Ireland, in County Down, in 2008. There are now more than 1,600 pairs in Britain.

SPARROWHAWK ■ *Accipiter nisus* Length 28–38cm; wingspan 55–70cm

DESCRIPTION The Sparrowhawk has short, broad wings and a long tail. The male is grey above, with pale underparts narrowly barred orange. The female is brown above and barred brown below. In flight, the barred underwings and tail are revealed. As well as being very different from the male in colouration, the female is 25% larger than the male and nearly twice as heavy, *c.* 260–140g.

HABITAT Principally a woodland bird, often moving to farmland hedgerows and gardens.

HABITS The Sparrowhawk flies with rapid burst of wingbeats, then a glide. It hunts by dashing from perch to perch, snatching prey in flight or from its perch – including from bird tables. The food consists almost entirely of birds. The male takes small songbirds, as does the female, but she can kill birds as big as a Woodpigeon or dove. Sparrowhawks are quite often seen being mobbed by Swallows or Starlings above their hunting grounds. The female lays 4–6 bluish-white eggs, streaked brown or reddish-brown, in a stick nest in a tree. Voice: a shrill, chattering *gek, gek*. Due to the birds' stealthy, low-level and speedy hunting technique many people miss it in their gardens, even though it may use the garden as part of its hunting route; it may sometimes be seen watching from a vantage point in a garden. You will know you have had one in the garden if you find a pile of pigeon feathers; the Wood Pigeon's distinctive grey-white-black tail feathers are a clue (see photo opposite below).

DISTRIBUTION Widespread across the British Isles throughout the year. It is a most abundant breeder in lowland central and eastern England. Organochlorine pesticides used on farmland in the 1960s, which persist in the food chain, resulted in a population decline – the hawks ate birds that had eaten contaminated seeds, so they were poisoned too. Since the pesticides were banned the species has recovered well.

1st winter

1st winter

KESTREL ■ *Falco tinnunculus* Length 32–35cm; wingspan: 71–80cm

Amber listed

DESCRIPTION The male Kestrel has chestnut upperparts with blackish spots, dark-streaked buff underparts, a bluish-grey head, rump and quite a long tail, the last with a black sub-terminal band and white tip. The female is reddish-brown above with dark barring, paler below with dark streaks, and has a brown, barred tail. The birds' wings are pointed.

HABITAT Occurs on farmland, hill country, moorland and rocky coasts.

HABITS Habitually hovers, facing into the wind, looking for prey, and suddenly dropping onto it – a common country name for it is 'Windhover'. It is the hovering bird often seen above the verges of main roads. Its diet consists especially of voles, and in warmer times, insects. It nests in old crows' nests, on cliff ledges and in hollow trees, large nest boxes and ruined buildings. The female lays 3–6 white eggs that are heavily marked with red-brown. Voice: commonly utters a clear *kee-kee-kee*.

DISTRIBUTION A mostly resident bird, the Kestrel rivals the Common Buzzard as Britain's most abundant and widespread daytime raptor in many places in summer and winter. It is most abundant in central and eastern England, and is more scarce in northern and western Scotland, Wales and south-west England. There has been an overall decline of the species of more than 40 per cent since 1970.

♀

PEREGRINE FALCON ■ *Falco peregrinus* Length 36–45cm (male), 46–51cm (female); wingspan 89–100cm (male), 104–113cm (female)

DESCRIPTION The Peregrine has broad-based, pointed wings and a tapering tail. The adult male (tiercel) has dark grey-blue upperparts and tail, and buffy-white underparts, finely spotted and barred black. The adult female (falcon) is darker and more heavily barred, and larger than the male. Juveniles are browner above and have streaked, not barred, underparts.

HABITAT Occurs chiefly in open country with coastal or inland cliffs, including quarries, and also man-made cliffs – tall buildings.

HABITS The Peregrine hunts by circling high, spotting prey below, then diving at incredible speed for the kill; it is one of the fastest birds in the world, with the dive's terminal velocity being up to 322kph. Its diet consists of birds, especially seabirds and pigeons, and it has recently been proved to hunt migrants such as waders at night. The female lays 3–4 buff or cream eggs with heavy red-brown markings, often on bare ground. Voice: utters a shrill *kek-kek-kek-kek*.

DISTRIBUTION The species is now widespread in the British Isles in upland and coastal areas with suitable breeding sites; it is less common in south-central and eastern England. There were marked declines worldwide in the mid-20th century due to poisons in the food chain. The species has recovered well in the British Isles, experiencing a more than 40 per cent increase in distribution in the last 40 years. It is even breeding successfully on big buildings in English cities, such as a church tower in Exeter, Chichester Cathedral and Westminster, London (more than 20 pairs nest in the city).

juvenile

MOORHEN ■ *Gallinula chloropus* Length 32–35cm; wingspan 50–55cm

DESCRIPTION The adult looks dull black at a distance, with a white line along its side; the white under the tail is very noticeable when it flicks its tail. It has a conspicuous red bill and basal shield, with a yellow tip, and yellowy-green legs and feet. Juveniles are dull dark brown, paler below.

HABITAT Favours still or slow-flowing streams, freshwater ponds and lakes, drainage ditches on farms and ponds in city parks.

HABITS Constantly flicks its tail as it swims or walks. To take off from land or water it has to run to gain speed to get 'lift', and its legs trail in flight. Its diet consists mostly of vegetable matter. It feeds on the water, and in the open on grassland. The female lays 5–9 buff eggs, variably marked with brown, black and purplish, in a nest of twigs and grasses in waterside herbage or even on a tree branch overhanging water. The species has 2–3 broods a year, and the earlier young help to feed new chicks. Voice: the common call is *curruc, curruc*.

DISTRIBUTION Resident and sedentary throughout Britain, but absent from high ground in Wales, the Lake District, and northern and western Scotland. The winter and summer distributions are very similar, and this is when the birds are most abundant in central and eastern England. Moorhens are most likely to be seen in a garden if it is near a source of water as described above, and where they are attracted to a liberal spreading of seed on the ground.

juvenile

LAPWING ■ *Vanellus vanellus* Length 28–31cm; wingspan 82–87cm

Red listed

DESCRIPTION The adult has a black-and-white patterned head and diagnostic crest, black breast, white belly and pale orange under-tail. The back is black, and there is metallic green on the mantle and wing-coverts. The wings are broad and rounded, and black with white tips, and the tail is black and white. Juveniles have a shorter crest and paler feather edges than adults.

HABITS Found mainly on open arable ground; also favours rushy fields and moorland for breeding. Hard weather drives flocks to muddy estuaries.

HABITS The male has a spectacular aerial display in spring, involving twisting, diving and calling. Winter flocks have a 'leisurely' flight with slow wingbeats. The diet consists mostly of a variety of invertebrates, taken from on or in the ground. Four olive- or clay-coloured eggs with black blotches are laid in a scrape on the ground. Voice: calls comprise variations on *PEE- wit*, which gives it its other common name, Peewit. Also sometimes known as Green Plover.

DISTRIBUTION Widespread across Britain, but has experienced a severe decline in the last 40 years because of changes in farming such as the sowing of winter rather than spring cereals, so that the crop is too tall to nest in in the following spring, less mixed farming and too much improved grassland resulting in less food. As a result of these factors, it no longer breeds in much of south-west England, western Wales and western mainland Scotland – although it still breeds on the western and northern isles. It is mainly a summer visitor to breeding sites; many individuals from western maritime mainland Europe winter in the British Isles. Lapwings are very susceptible to cold weather, which makes them move on; this makes it possible to see a flock flying south or west over a garden.

BLACK-HEADED GULL ■ *Chroicocephalus ridibundus*

Length 34–37cm; wingspan 100–110cm

Amber listed

DESCRIPTION The summer adult has a chocolate-brown (not black) hood, pale blue-grey upperparts, and a white body and tail. The wings are tipped black, with a diagnostic broad white band at the front of the primaries, the undersides of the primaries are dark grey, and the bill and legs are red. The winter adult has no hood, just a dark spot behind the eye. Juveniles have a yellowish bill and legs, black trailing edges to the wings behind a white panel, and a black-tipped tail.

HABITAT Breeds on sand-dunes, and beside lakes and marshes, in a wide range of coastal and inland, natural and man-made wetlands. Winters on estuaries, playing fields, lakes in parks and reservoirs.

HABITS The diet consists especially of insects and worms, and Black-headed Gulls are often seen following the plough. In many places, near the coast especially, they land in gardens where scraps have been thrown out. The species is colonial at nest sites, and is gregarious in winter, often occurring in very large flocks. A simple nest of a pile of vegetation is made by both birds on the ground, in a site chosen by the male. The female lays 2–3 greenish or brown eggs with darker spots. Voice: utters a distinctive laughing call.

DISTRIBUTION About 200,000 pairs nest in Britain; about 3,000,000 also winter in Britain, coming from Scandinavia and eastern Europe. The species is most common as a breeder in eastern Scotland, and northern and south-east England. It is scarce in Wales and southern England, and hardly any individuals nest in south-west England – the estuaries there are empty of these gulls in summer, while they are present in their hundreds in winter.

adult summer

adult winter

1st winter

COMMON GULL ■ *Larus canus* Length 40–47cm; wingspan 110–130cm

Amber listed

DESCRIPTION This gull has a smallish bill and rounded head. The summer adult has a white head and body, and darkish grey wings with black primaries and large white tips (mirrors). Bill and legs yellowy green. Immatures have almost black primaries, pale coverts and a blackish bar across the trailing edges of the secondaries; the white tail has a black subterminal band. Adult plumage is achieved in the third year.

HABITAT Nests by freshwater lochs, on moorland, inshore inlets and marshes. After breeding, occurs on estuaries, grassland and sports fields.

HABITS Common Gulls are colonial nesters that are sociable outside the breeding season, often feeding or roosting with other gulls, which is when the garden birdwatcher is most likely to see one; it is best to listen for its distinctive call. The species' diet consists of fish, and marine and land invertebrates. Three olive or blue-green eggs with brown markings are laid in a nest of vegetation or seaweed on the ground. Voice: the call is a high-pitched *keee-ya*.

DISTRIBUTION Common Gulls winter widely in Britain, except in the highlands of south-west England, Wales, Cumbria and Scotland. At this time they are most abundant in lowland eastern Scotland southwards, with the numbers being swollen by many immigrants from continental Europe; they are least common in Wales and south-west England. The breeding distribution is very different: almost all the colonies are in Scotland, mostly north of the Southern Uplands, especially from Angus to the Moray Firth, and around the coast of Northern Ireland. There are only a handful of small colonies scattered in England. This is not the most common gull, despite its English name – its breeding population is only about a third of the size of that of the Black-headed Gull (see p. 45).

adult winter

adult summer

1st summer

1st winter

LESSER BLACK-BACKED GULL ■ *Larus fuscus*
Length 48–56cm; wingspan 117–134cm

Amber listed

DESCRIPTION The adult has a white head, body, tail and underwings. The mantle and upperwings are dark grey, the wing-tips are black with white primary tips, and the legs and feet are yellow. Juveniles have a streaked grey-brown body and wings. At all ages and plumages, care should be taken to separate this species from the Great Black-backed Gull (see pp. 52–53). Adult plumage is achieved in the fourth year.

HABITAT Nests on coastal and lake islets. Outside the breeding season occurs on estuaries, refuse tips and playing fields.

HABITS This species is omnivorous, even feeding on refuse. It is a colonial ground-nesting breeder, but in recent years some nests have been built on the rooftops of coastal cities. The female lays 2–3 eggs with a variable ground colour, heavily marked with blackish-brown. Voice: the calls consist of *kaw* and *ga-ga-ga*, similar to the calls of the Herring Gull (see pp. 50–51).

DISTRIBUTION Many but not all UK birds are migrants, returning to their breeding grounds around Britain's coasts in late February or early March, and leaving to winter as far south as the coast of Morocco. The species increasingly winters in England, often far inland. By fixing small tracking devices to birds from Suffolk, it was deduced that birds from the same colony did not all winter together; one female wintered in Portugal, whereas her mate went only as far as Dorset and Hampshire. Another bird flew for three years across the Bay of Biscay and around the coast to Gibraltar, but returned each time to the French coast and beyond by crossing Spain from the Costa del Sol via Madrid. This species is not a regular garden scrounger like the Herring Gull, and is more likely to be seen flying overhead.

juvenile

HERRING GULL ■ *Larus argentatus* Length 55–67cm; wingspan 138–155cm

DESCRIPTION The adult's upperparts are pale grey, with white leading and trailing edges to the wings; the wing-tips are black with separated white 'mirrors', and the rest of the plumage is white. First-winter birds are dusky mottled brown with dark brown wing-tips and tail tip. Adult plumage is achieved in the fifth year.

HABITAT This is *the* typical seagull. It nests on coastal cliffs and islands, and has recently been nesting on buildings in coastal cities and towns, sometimes a few kilometres from the sea. Non-breeders can be seen on estuaries, refuse dumps and fish quays, and following ships not far offshore.

HABITS Herring Gulls are omnivorous and will eat almost anything. The female lays 2–3 olive eggs with many dark brown spots and blotches, in a mound of vegetation constructed by both sexes. Herring Gulls are very bold: they may arrive unexpectedly in a garden to grab scraps that have been thrown out, without the owner noticing that any gulls were cruising overhead, searching (this behaviour also occurs in Black-headed Gulls). The scraps should not be overdone – some people think they are helping by throwing out whole slices of bread, but much of such food becomes rubbish and attracts rats. Seaside resorts are trying to discourage visitors from throwing food for gulls, since this encourages the birds to come close, when they may steal food like fish and chips and even ice-cream from the hands of seaside tourists. Voice: the call is a loud *kyou-kyou-kyou*.

DISTRIBUTION Herring Gulls are widespread. Unlike Black-headed and Common Gulls, this common breeder nests almost exclusively on coastal sites all around Britain. It

has, however, colonized certain urban areas, some of which are several kilometres from the coast. It has declined in numbers in recent years – there have been fewer refuse sites and other sources of food to scavenge, and due to lack of food many colonies are now reduced in size. The species can be seen widely in winter in the UK, except in highland areas. Far northern and eastern populations migrate south and west; others disperse to local estuaries and ports.

2nd winter

1st winter

GREAT BLACK-BACKED GULL ▪ *Larus marinus*
Length 64–78cm; wingspan 150–165cm

Amber listed

DESCRIPTION The adult bird is very large – it is the biggest gull nesting in Britain, with an all-black back and wings, the latter with white tips and trailing edges. The rest of the plumage is white, and the legs and feet are pale pink (compare with Lesser Black-backed Gull, pp. 48–49). Immatures are mottled and streaked brown, but with a paler head. Adult plumage is achieved in the fifth year.

HABITAT Nests on small islands and stacks. In winter it scavenges at ports and refuse dumps. It is coastal in winter.

HABITS This gull has an omnivorous diet. It is a serious breeding season predator of shearwaters and Puffins, catching many of these birds in mid-air. It nests singly or in a small group. The female lays 2–3 olive-brown eggs blotched darker brown on a 'nest' – actually just a pile of vegetation – on the ground. Voice: utters a deep, barking *aouk, aouk*.

DISTRIBUTION The Great Black-backed Gull breeds around British coasts, except the North Sea coast from Lothian in Scotland to East Anglia in south-east England. By contrast, it winters widely around British coasts, and also well inland except the highlands. In summer non-breeding birds are seen well inland, too (the birds do not mature until they are five years old). Most birds are resident, simply dispersing to winter along Britain's coasts and offshore. Far north-eastern populations migrate west and south. This species is most likely to be seen from a garden while flying over in winter from one food source to another.

1st winter

WOODPIGEON ■ *Columba palumbus* Length 40–42cm; wingspan 75–80cm

DESCRIPTION This is Britain's largest pigeon. Many pre-1950s bird books call it the Ring Dove or Wood Pigeon. It has a grey head and upperparts, a green-and-purple sheen on its neck, and a pink breast. An adult can be identified by the white patch on each side of its neck (hence the species' old common name). It has a unique white bar across (not along) the wing-coverts, and a grey tail with a black terminal band above and grey, white and black below (a very distinctive feather often found on a garden's lawn). Juveniles lack the neck-patches.

HABITAT Occurs in wooded country bordered by fields, and parks and gardens. There has been a big increase in the species in parks and cities in the past 40 years.

HABITS Woodpigeons have a distinctive, deeply undulating display flight, with wing claps at the top of each rise. The diet comprises weed and crop seeds, and the birds are fond of seed put out on bird tables, which they quickly clear (tits and finches must thus be fed from hanging feeders). Woodpigeons need fresh water for drinking, as do other pigeons. Two white eggs are laid in a very basic nest platform of twigs in a bush or tree. Voice: the birds' *coo COOO coo coo-coo-cuk* can be heard during much of the year. To many people the bowing, displaying males are an attractive feature of garden birdwatching.

DISTRIBUTION Woodpigeons are widespread throughout Britain in summer and winter, except highland and north-west Scotland and the outer isles. Most British birds are resident, but winter populations are swollen by hundreds of migrants from mainland Europe, at which time Woodpigeons can form very large flocks. They are abundant breeders in arable eastern lowlands. BTO studies in particular have shown that the population increased by 169 per cent in 1967–2010. So many Woodpigeons visit gardens now, even in city centres, that many garden birdwatchers are having to devise ways of feeding small birds separately so that the pigeons do not gobble up all the seeds on bird tables.

STOCK DOVE ■ *Columba oenas* Length 32–34cm; wingspan 63–69cm

Amber listed

DESCRIPTION This is a mostly blue-grey bird. The broad, blackish tips and trailing edges to the wings show above and below; there are two narrow black bars on the greater coverts and a black band at the end of the tail. The neck-sides are glossy green, and the breast is vinous pink. The legs are bright pink, and the bill is horn coloured with a white cere.

HABITAT Found in copses, ancient parkland and city parks – mostly everywhere where there are trees close to open ground.

HABITS The Stock Dove feeds on plant material, especially seeds on lowland arable farmland with open woodland. Two white eggs are laid in a nest in a tree-hole, rock crevice, ruined building or large nest box. Voice: the song is repeated quickly up to ten times, *Oooo-er*.

DISTRIBUTION The Stock Dove is widespread in the British Isles, but is the least common and least often seen of the four doves in this book. It is largely sedentary, and is found widely in England, Wales, and southern and eastern Scotland, and not at all in the rest of Scotland; it is scarce in Northern Ireland. The species is never as bold as the Woodpigeon, so is most likely to be seen in a garden only if it has mature trees and is close to farmland.

ROCK DOVE or FERAL PIGEON ■ *Columba livia*
Length 31–34cm; wingspan 63–70cm

DESCRIPTION The Rock Dove is the ancestor of all domestic pigeons, including feral pigeons. It is one of four birds described here that are all from the same family, but are called either 'dove' or 'pigeon'. The former tend to be smaller birds, but there is no scientific difference between them. Rock Dove adults are grey, with a darker head, breast, flight feathers and terminal tail-bar. Two black bars cross the secondaries. There is a green-and-purple sheen on the neck, and the white rump shows well in flight. Some feral pigeons in urban areas are similar, but many have mottled or speckled plumage, often with a considerable amount of white and/or brown.

HABITAT Rock Doves traditionally inhabit cliffs and rocky coastal landscapes, but see below. They nest in colonies (hence the ease with which the species was tamed). They are resident across Britain except in highlands. The best places to find birds in wild plumage are far northern and western Scotland and western Ireland. Their descendants are found most abundantly in central and eastern Britain, breeding in small colonies in urban areas on large buildings and under road and rail bridges, because these structures mimic the traditional cliffs.

HABITS The diet consists of seeds, green leaves and insects. The birds visit town gardens where they discover regular handouts of bird seed. Two white eggs are laid in a simple nest on a ledge or crevice. Voice: the song is *oor-roo-coo*.

DISTRIBUTION It is very hard to know the original distribution of the Rock Dove, because feral birds are so common throughout Britain.

COLLARED DOVE ▪ *Streptopelia decaocto* Length 31–33cm; wingspan 47–55cm

DESCRIPTION This dainty dove is the smallest of the four species described here. The adult is mostly pale pinkish-buff with dusky-brown primaries. The upper tail is brown with a white tip, and especially broad at the corners; it is white with a black base below, and very noticeable in flight. There is a thin black bar outlined with white at the sides of the neck. Juveniles lack the neck-bar.

HABITAT Collared Doves occur near human habitation, especially farms, and gardens with bird tables.

HABITS In display flight, Collared Doves rise steeply then glide down, often calling a harsh *kurr* as they land. Their diet comprises grain and weed seeds. The female lays two white eggs in a delicate, thin structure of twigs in a bush or tree, through which the eggs can sometimes be seen from below. Voice: sings repeatedly from an open perch, *coo-OO-cuk, coo-OO-cuk, coo-OO-cuk, coo-OO-cuk*. Many people welcome these dainty doves, but some are driven mad by the incessant cooing.

DISTRIBUTION There has been an amazing spread of Collared Doves from the Balkans since the 1930s. The birds first bred in Sweden by 1951, and in south-east England by 1955. They are widespread throughout the British Isles, from the Orkneys to the Scilly Isles, with very similar breeding and winter distributions. They are mostly resident; juveniles disperse. The reasons for their explosive spread are not fully understood, but it may be that they exploited the spilt grain and livestock food on farms, then benefited from the bird food available in gardens. They are not seen in highland areas, and are most abundant in eastern England.

CUCKOO ■ *Cuculus canorus* Length 32–34cm; wingspan 55–60cm

Red listed

DESCRIPTION A bird with a small head, decurved bill, slim body and long tail, the Cuckoo has a slate-grey head, breast and upperparts, and darker, pointed wing-tips. The rest of the underparts are white barred with brownish-black bars. The white-tipped tail droops below the wings at rest. The female is buff on the lower breast (the rare rufous phase is barred with black). Juveniles have a plain grey phase and a rufous phase, both with a white patch on the nape.

HABITAT Occurs in a wide range of habitats, but mostly in uplands like Dartmoor, Exmoor and western and northern Scotland, marginal habitats such as reedbeds, and heaths like those in the New Forest.

HABITS An infamous and promiscuous parasitic nester, the female Cuckoo lays up to 25 eggs, with one per nest; the Reed Warbler, Dunnock and Meadow Pipit are the main hosts. The Cuckoo flies with its wings held below shoulder level, and is often mobbed by small birds. The male's call, the famous *cuc–oo*, is often repeated several dozen times, one after the other. The female has a bubbling call.

DISTRIBUTION The Cuckoo is a summer visitor across the region, even the Arctic, but is not found in open tundra. It is widespread in Britain except in urban areas, but has suffered a worrying decline throughout much of Britain of over 50 per cent since the mid-1990s. It winters in tropical Africa, mostly south of the Equator, where birds have been found as a result of modern tracking methods (for information, see http://www.bto.org/science/migration/tracking-studies/cuckoo-tracking). The species is most likely to occur in gardens if they are rural and close to its optimum habitat. Reporting it to the local bird club or the BTO (see p. 20) helps build that year's report on how widespread and common the species is.

juvenile

RING-NECKED PARAKEET ■ *Psittacula krameri*

Length 38-42cm, of which 25cm is tail; wingspan 42–48cm

DESCRIPTION An unmistakable, nearly all pale green bird with a long, pointed tail that has a bluish sheen. The bill is large, crimson and hooked. The male has a bluish sheen on the nape, and a black-and-pink ring around the neck. Juveniles are like the female, but with shorter tails.

HABITAT Found in gardens, parks and orchards. Noisy flocks of hundreds of non-breeding birds come to roost in favourite mature trees in south-east England.

HABITS This is a strong flier with long tail streaming. It feeds on fruits on trees, wild berries and seeds at bird tables. It nests in an existing hole in a tree as a solitary pair or in a colony, with each pair laying 3–4 eggs. Voice: utters loud screeches from a perch or in flight.

DISTRIBUTION In Britain Ring-necked Parakeets first bred in Kent in 1971; they may have been escapes from captivity, or deliberately released. They are native to India. The species is increasing in Britain, with a scatter of records throughout southern, central and northern England. It is mostly resident in south-east England, but has been seen as far as south-west England and Scotland. It is more likely to wander in winter than in summer, so there are a greater number of English records then, except in the south-west. There is an abundant concentration of the birds in the Greater London area and Kent, now getting on for 9,000 breeding pairs, and many more non-breeding birds.

BARN OWL ■ *Tyto alba* Length 33–35cm; wingspan 85–93cm

Amber listed

DESCRIPTION British and French birds have mottled grey-and-buff upperparts and wing-coverts, and pure white underparts; eastern birds have mainly grey upperparts and yellowish-orange underparts, finely speckled with brown. In flight the underwing is noticeably white, and the flight feathers are buff with faint darker bars. The heart-shaped facial mask is white, thinly outlined with dark brown, and the eyes are black.

HABITAT Occurs on farmland, and in rough grassland, hedges and copses.

HABITS At dusk the Barn Owl looks ghostly, quartering an area of open ground silently, slowly and low down, and often hovering. It is crepuscular and nocturnal. It feeds on small mammals. The female lays 4–7 white eggs on the bare floor, on a ledge in a barn or church tower, or in a hole in a tree; in recent years nest boxes have been used. Voice: utters a drawn-out *shreeee*, hence its old name, Screech Owl.

DISTRIBUTION Absent from urban areas, much of Northern Ireland, high altitudes and the Scottish islands. It has been in decline in many countries since the mid-19th century due to a range of factors, including changes in agricultural practices, cold winters, road-traffic deaths because many birds hunt over roadside verges, and loss of nest sites (barns have been renovated or turned into homes or demolished). A resident species, it is most abundant in eastern England and East Anglia. In several areas conservation has benefited the population through the provision of special nest boxes.

TAWNY OWL ■ *Strix aluco* Length 37–39cm; wingspan 94–104cm

DESCRIPTION The Tawny Owl has a large, round head, broad, rounded wings and a greyish-brown face-mask. Most of the plumage is rufous-brown in western European birds, and greyer in eastern European birds, streaked above and below; the wings have two broken white bars.

HABITAT Found in woods, copses, parkland and even town suburbs, nesting in tree-holes, old nests and specially constructed nest boxes.

HABITS Nocturnal, and roosts in thick cover. It is mobbed if found by small birds. Tawny Owls hunt at night on silent wings, mainly for rodents, and also birds and insects.

Indigestible bones and fur are ejected in a pellet. The female lays 2–5 white eggs on the bare floor of the nest-hole, which is in a tree. Voice: the male calls *hoo-hoo*, and both birds call *ke-wick*. The two calls together form the traditional but inaccurate *tu-whit-to-woo*, and the birds are most vocal in autumn.

DISTRIBUTION Resident, and evenly distributed throughout Britain in woodland, but absent from Northern Ireland, highlands, Scottish islands and the Isles of Scilly. Found even in cities if mature trees are part of the habitat in parks and gardens. Because of the Tawny Owl's nocturnal habits, it is generally most likely to be heard rather than seen.

juvenile

LITTLE OWL ■ *Athene noctua* Length 21–23cm; wingspan 54-58cm

DESCRIPTION This is Britain's smallest owl, the size of a Starling. Despite its small size it always looks fierce because of its black-and-yellow eyes staring out under white eyebrows. The greyish-brown upper plumage is liberally spotted with white, and the pale breast is well streaked with brown. Proportionally, the tail is short and the wings are quite long.

HABITAT Mostly occurs on lowland farmland with woods, copses and hedges with old trees, in England and Wales.

HABITS Little Owls are seen quite often in daylight, perched erect and motionless on a bare branch, telegraph pole or rock. When alarmed they have a habit of bobbing. The diet consists of insects, earthworms, small birds and amphibians, and Little Owls regularly run or hop on the ground when hunting for invertebrates. The female lays 3–4 white eggs, usually in a tree-hole, but buildings and cliffs may also be used for this purpose. Voice: most frequently heard in spring, a sharp *kiew, kiew*.

DISTRIBUTION A widely found but uncommon sedentary resident. It is largely restricted to England, where it was introduced in the 19th century; two schemes, in Kent and Northamptonshire, were successful, and it slowly spread across southern England and northwards, but it is found only sparsely in south-west England and there are very few birds in Wales. It is most abundant in mixed habitats. Changes in farming practices, reducing its food supply, are thought to have reduced its numbers.

COMMON SWIFT ■ *Apus apus* Length 16–17cm; wingspan 42–48cm

Amber listed

DESCRIPTION This is an all blackish-brown bird with a not very noticeable whitish throat. The long, narrow wings are scythe shaped, and the tail is short and forked. It is sometimes misidentified as a Swallow (see pp. 90–91), but that bird has extensive white underparts.

HABITAT An aerial species, nesting in crevices in cliffs and crags, and buildings in villages and towns.

HABITS Swifts employ a fast, agile flight, often seeming to have alternate wingbeats. They feed on insects caught in flight, catching several thousand a day in the breeding season. The birds wander widely, often for many kilometres, to hunt insects in the air, so they are seen over a wide variety of habitats. Their natural nesting habitat has been almost wholly replaced by buildings; conservationists are asking homeowners to put special nest boxes in their modern roofs, which do not have the gaps under the eaves that would lead to nest cavities. Swifts are very noticeable birds at their nest sites, where several pairs may race around screaming a shrill *sreeeeee*. The female lays 2–3 white eggs in a nest made of grass and other vegetable material, and feathers collected in flight.

DISTRIBUTION Summer visitor, late April to August. Winters in southern Africa. Widespread but only thinly present as a breeding bird in Cumbria, northern and western Scotland, and the Scottish islands – it is most likely to be seen there only on migration. It has declined by nearly 40 per cent since the mid-1990s, due it is believed to the lack of nest sites on modern buildings, inclement summer weather reducing the food supply and changes in conditions in its winter range in southern Africa.

GREEN WOODPECKER

■ *Picus viridis* Length 31–33cm; wingspan 40–42cm

Amber listed

DESCRIPTION Adult Green Woodpeckers are yellowy-green above and pale greeny-grey below, with lightly barred flanks; in flight, they show a bright yellow rump and cream bars across the dark brown primaries. The head pattern is distinctive: there is a red crown and nape, black around the eye, and a black moustache, the male's with a red centre. Juveniles have white-spotted upperparts and brownish-black bars on the face and underparts.

HABITAT Favours deciduous woodland, orchards and parkland, and seldom wanders far from its home territory.

HABITS This is a very specialized woodpecker, feeding almost exclusively on ants and their pupae, mostly on the ground, digging into ants' nests and extending its very long, sticky tongue into them. It is most likely to be seen in gardens that have active ants' nests. The birds nest in specially excavated holes – on average 28cm deep – in mature trees. The female lays 5–7 white eggs. Voice: utters a laughing *cue-cue-cue*, hence its old country name of Yaffle. This species drums only rarely.

DISTRIBUTION These are mostly resident birds, so the summer and winter distributions are virtually the same. They are found throughout England, Wales (except the west), and southern Scotland, except on high ground and the islands. They are most common in the south and east, become less frequent further north and west, and are not found in Northern Ireland.

juvenile

GREAT SPOTTED WOODPECKER ■ *Dendrocopos major*
Length 22–23cm; wingspan 34–39cm

DESCRIPTION This species has black upperparts with white scapulars forming big ovals, four narrow white wing-bars, and white underparts except for the crimson-red vent. It has a whitish forehead, black crown (the rear of the male's crown is red), and a black moustache from the base of the bill around the white cheeks, meeting at the black nape; the bill is black.

HABITAT Occurs in all kinds of woodland.

HABITS This woodpecker uses its stiff tail to help control its climbs up trees, and it has a bouncing flight. Its diet comprises mainly insects and larvae. It eats many coniferous seeds in winter; if seed sources fail in the north, numerous birds erupt in the south and west. The birds excavate nest-holes about 28cm deep, and over several years they may drill one hole above the other in the same tree. The female lays 4–7 white eggs. Voice: the call is a sharp, clear *kik*, which carries well in woodland. It drums on a dead branch with its bill in spring for about 0.5 seconds, the drumming dying away at the end.

DISTRIBUTION Although this bird is mostly resident, juveniles wander, and all may do so if their natural food is in short supply. Much more widespread than the Green Woodpecker, this species breeds throughout Britain but is scarce in Northern Ireland; it does not breed on the Scottish islands and Scilly Isles, but may sometimes be seen there as a migrant. It is most common in south-east England and the counties bordering Wales. Since the late 1960s there has been a four-fold increase in the population, with a related range expansion through Scotland, and to the far west of Wales and east of England. The birds are increasingly coming to bird tables, and are teaching their young where to find easy food by bringing them, too.

juvenile (left), female (right)

JACKDAW ▪ *Corvus monedula* Length 30–34cm; wingspan 64–73cm

DESCRIPTION This is a small black crow with a short bill; at a distance it looks all black, but at closer range the grey rear to the back of the head can be seen; this is like a grey hood, with clear-cut edges between the crown and mantle. The eyes are shiny and pearly white. The Jackdaw employs faster wingbeats than the Rook or other crows, so with practice it can be easily identified in flight.

HABITAT Breeds and winters in wide range of habitats: old woodland, parks, coastal cliffs, quarries and urban areas, with some open ground on which to feed.

HABITS The Jackdaw is an agile flier, especially across cliff faces and tall buildings, appearing to play in the air currents, tumbling and diving. The species pairs for life, so the birds are often seen in close company. Their diet comprises invertebrates, seeds, fruits and carrion. As a ground feeder the Jackdaw is very dependent on pastureland, often foraging with Rooks and Starlings, but it happily feeds in a garden if it finds a regular food supply there. It nests in small colonies in tree-holes and crevices in cliffs, and even close to humans in nest boxes, holes in church towers and other buildings, and down chimneys. The birds often make what seem to us to be comic efforts to try and get a long stick, carried crosswise in the bill, through the nest-hole. The nest is constructed from sticks and lined with wool and hair; 4–6 light blue-green eggs marked with brown are produced. In the non-breeding season the birds roost in trees in large numbers. Voice: the call is a sharp *chack*, often repeated, along with other quieter, more chattery, cawing and *kyaar* calls. The species has a most unusual name because both parts – jack and daw – are onomatopeic.

DISTRIBUTION The Jackdaw is mostly resident throughout Britain, in summer and winter, except in north-west Scotland and the Outer Isles. The British population has grown by more than 100 per cent since the 1970s.

ROOK ▪ *Corvus frugilegus* Length 41–49; wingspan 81–94cm

DESCRIPTION The Rook is an all-black bird glossed with green and purple. It has a sharp, pointed black bill with whitish-grey at the base, a steep forehead, and a rounded, almost wedge-shaped tail-tip. The bill, head and tail shapes distinguish it from other crows. The loose thigh feathers give the bird a 'baggy-trousered' effect. Juveniles have black at the base of the bill.

HABITAT Mostly found in agricultural country with some tall trees for nesting colonially (in a rookery).

HABITS Rooks are gregarious outside the breeding season for feeding and roosting. The diet consists especially of earthworms, invertebrates and grain. Rookeries may contain hundreds of tree-top nests; these are often close together, and are mended or rebuilt each year before the trees are in leaf. There may be thousands of birds in a post-breeding roost, and their evening flights to the roost are spectacular. The female lays 2–5 blue-green eggs well marked with brown. Voice: the call is *kaah*, less harsh than a Carrion Crow's. If you live near a rookery, be prepared for non-stop, nearly deafening sound, especially during nest building (plenty of stick stealing results in aggression between birds) and mating.

DISTRIBUTION Resident in Britain except in large cities, north-west Scotland and highlands in Wales and northern England. Rooks are largely sedentary, favouring open countryside for ground feeding, and copses for nesting and roosting.

CARRION CROW ■ *Corvus corone* Length 44–51cm; wingspan 84–100cm

DESCRIPTION This crow is all black, with no white at the base of the bill. It has a less steep forehead and more square-ended tail than the Rook (see p. 71), and less of the 'baggy-trousered' look of that bird.

HABITAT Found in a wide variety of habitats – by the coast on cliffs, arable land, grassland, parkland, heaths and moors, estuary mud and refuse tips, and in cities.

HABITS The Carrion Crow feeds on the ground, walking with a steady gait. It eats almost anything. It is not as gregarious as the Rook and is often seen singly, but small flocks form at good food sources such as refuse tips, and estuary mud or sand, where they look for shellfish. The birds nest in single pairs, not in a colony like its relative, usually in trees, constructing a large stick nest with a soft lining. The female lays 4–6 light blue or green eggs liberally marked with brown spots. Carrion Crows do nest in city centres if nest sites and food are available. Voice: the call is a hoarse *kraah*.

DISTRIBUTION This is a widespread and common resident in England, Wales, and southern and eastern Scotland. It is most common in lowland farmland. Hybrids occur at the borders of the distribution of this and the Hooded Crow (see opposite). The Carrion Crow wanders more in winter than in summer, even reaching the Northern Isles and western Scotland at this time.

HOODED CROW ▪ *Corvus cornix* Length 44–51cm; wingspan 84–100cm

DESCRIPTION Some authorities still believe that this is a subspecies of the Carrion Crow (see opposite). The 'Hoodie' has a dirty-grey body between a black head and breast, and black wings and tail. Hybrids show great variations in colour patterns between the two species; some are largely black, others have just a little mottling on the grey.

HABITAT Found in a wide variety of habitats – by the coast on cliffs, and in open woodland, parkland, heaths and moors, and even towns.

HABITS Like its relative, the Hooded Crow is omnivorous. It nests in pairs in the crowns of trees, laying 4–6 eggs very like those of the Carrion Crow. Voice: the call is a hoarse *kraah*.

DISTRIBUTION This is a resident species that breeds in Ireland, the Isle of Man, and north-west Scotland and its islands. It replaces the Carrion Crow in north-west Scotland from Argyll to Sutherland. Hybrids occur between the two species in the overlap zone and on the Isle of Man, where both occur. Some birds wander in winter: they are then recorded in England, especially on the east coast, although these birds may be migrants from eastern Europe or Scandinavia. Interestingly, the Hooded Crow is found throughout Ireland where there are no Carrion Crows, apart from some pairs in the north-east.

RAVEN ■ *Corvus corax* Length 54–67cm; Wingspan 120–150cm

DESCRIPTION The Raven is the largest crow, half as big again as a Carrion Crow or Rook. It is all black with a wedge-shaped tail, massive bill, flat head and long wings. Its shaggy throat feathers are particularly noticeable when it is calling. In flight the heavy head and bill help to give the bird a cruciform silhouette.

HABITAT Occurs from sea level to high mountains, usually avoiding forest interiors and intensively farmed land. It needs undisturbed nest sites on sea cliffs, or in quarries or trees. The Raven has become a town bird again in some places – a scavenger as of old, even nesting in city parks.

HABITS The Raven has a powerful flight, soars freely and performs aerobatics (diving and spectacularly flipping onto its back), mainly in the breeding season. It ranges widely for a variety of food, especially carrion. It is mostly seen singly or in pairs, but small groups form at important food sources such as a dead sheep. Its large stick nest is reused, often for several years. The female lays 3–7 light blue or green eggs heavily marked with brown. Voice: the common call, often voiced in flight, is a repeated, deep *pruk* or *kronk, the* atmospheric sound of wild country.

DISTRIBUTION Ravens are sedentary, although juveniles disperse. They were formerly widespread, but are now most common in northern and western Britain, and not as confined to wild uplands as they were in the 20th century, after they had been persecuted to extinction in eastern Britain in the years before. At that time they were birds only of Britain's western uplands in Scotland, north-west and south-west England, and Wales. There has been a more than 60 per cent increase in the Raven's range eastwards since the late 1960s, where it breeds and winters widely, except in an area to the east of a line drawn roughly from Inverness to Kent, where it is absent or rare.

MAGPIE ■ *Pica pica* Length 40–51cm, of which 20–30cm is tail; wingspan 52–60cm

DESCRIPTION The Magpie is an unmistakable bird: the scapulars, outer half of the wings and flanks are white, with the remaining parts being black with a purple-and-green iridescence. The distinctive pied pattern shows in flight, as does the long, streaming tail. The sexes are similar. Juveniles have shorter tails and duller plumage than adults.

HABITAT Mainly a lowland bird in lightly wooded country; as numbers have increased it has moved into suburban and urban habitats in several countries.

HABITS Feeds on the ground, consuming invertebrates in summer, and vertebrates and seed in winter; also eats carrion and scraps. Constructs a distinctive stick nest with an unusual canopy of thorny twigs, most frequently in a tall tree. It is often easy to spot when both birds start to build it as early as February. The female lays 5–7 greenish eggs with dense mottling of brown and grey. Recent research has shown that the breeding success of small birds has not been harmed by the increase in numbers of Magpies. Voice: the common call is a loud, staccato *chacker chacker chacker*.

DISTRIBUTION The Magpie is a sedentary resident in Britain, excluding the Scottish Highlands and islands. Adults may spend all their lives in the same territory; first-year birds disperse to find their own territories. Overall there has been a nearly 100 per cent increase in numbers since the 1970s. Interestingly, the Magpie is most abundant in urban and suburban areas in a line from south-east England, to the Midlands, to Lancashire and Yorkshire. It is least abundant in upland areas.

JAY ▪ *Garrulus glandarius* Length 32–35cm; wingspan 54–58cm

DESCRIPTION The Jay is Britain's most colourful crow. It has a pinkish-brown body with a white rump and under-tail coverts, a black tail, a white forehead and crown streaked black, and a broad black moustachial stripe. The wings are black with a short white bar and blue-and-black flash. The sexes are alike, and juveniles are similar to adults.
HABITAT British birds are sedentary, occurring in the fairly dense cover of usually broadleaved trees.
HABITS The first view of a Jay may be as it is flushed from cover and flies away, showing the characteristic black tail and white rump. Jays eat invertebrates, fruits and seeds. Acorns are their staple winter diet; they come into suburban gardens in autumn in particular, seeking acorns (if there are oaks nearby), or to bury them in lawns or flower beds to provide food later. European populations are eruptive migrants when the acorn crop fails. The nest is a shallow, twiggy cup built in the fork of a tree. The female lays 3–10 smooth, pale blue-green or olive eggs with buff-coloured speckles. A shy bird, easily missed, despite its colours – yes, it really is a crow! Voice: the call is *skaaak skaaak*; it is loud, harsh, raucous and far carrying.
DISTRIBUTION In Britain the Jay is found from the Great Glen of Scotland southwards; it is absent only from highlands and lowlands where there is a lack of woodland, its preferred habitat. It has increased in Britain, and has even moved into parks in towns and cities, especially where there are mature trees.

GOLDCREST ■ *Regulus regulus* Length 9cm; wingspan 13.5–15.5cm

DESCRIPTION The Goldcrest is a tiny bird, weighing only 4.5–7.5g, with dull greenish upperparts, two white wing-bars and pale olive-green underparts that are darker on the flanks. It appears large-eyed, with black eyes on a whitish face. The crown in both sexes is yellow lined with black, and a displaying male raises his crest to reveal the orange centre. Juveniles lack the crest.

HABITAT Usually breeds in well-grown conifers, but will inhabit deciduous woods, gardens, cemeteries and parks where there are suitable conifers.

HABITS The Goldcrest is very active and always on the move, even hovering to pick insects off the undersides of leaves. It is insectivorous throughout the year, so suffers severely in bad winters. Censuses have shown that populations can recover well. The nest, in which 9–11 tiny eggs are laid, is suspended at the end of a conifer branch. Voice: the calls and song are too high pitched for some people to hear. The bird's tiny size, restlessness and high-pitched voice often make it difficult to see.

DISTRIBUTION Goldcrests are mainly resident, breeding in suitable habitat throughout Britain. They are most abundant in heavily forested areas, but visit and even nest in large gardens in suburbia. They are least abundant in intensively farmed land. The species is widespread in winter as well as at other times of the year, with birds often joining tit flocks, when they will wander from the breeding area to thick hedges and gardens with mature shrubs and trees.

FIRECREST ■ *Regulus ignicapillus* Length 9cm; wingspan 13–16cm

Amber listed

DESCRIPTION The Firecrest is tiny, weighing only 4–6.5g, and looks like a more brightly coloured Goldcrest (see p. 79). The diagnostic patterned striped head comprises a golden crown lined on each side with black, and a broad white supercilium underlined by a black eye-stripe; the displaying male raises his crown to reveal the orange 'fire-crest'; the female's crown is yellow. Both sexes have a bronze-coloured patch on each side of the neck. The upperparts are bright olive-green, and there are two white wing-bars; the tail is darker and browner, and all the underparts are white.

HABITAT Spends more time in broadleaved trees and shrubs than does the Goldcrest; in conifers it feeds in less dense branches.

HABITS The species prefers larger prey than its relative – arthropods, aphids, spiders and insect larvae. It breeds mostly in Norway Spruce plantations, in a nest similar to the Goldcrest's in which 7–12 pink eggs with some darker markings are laid. Voice: the call is a very high-pitched *zit zit*, lower than the Goldcrest's; the song is a rapid string of calls.

DISTRIBUTION The European Firecrest population has spread westwards. The species first bred in England in 1962. It now breeds mainly in southern and south-east England, and is rarely seen elsewhere in the breeding season. It is more widespread in winter, when migrants from colder mainland Europe arrive in the autumn. At this time it is most likely to be seen in gardens as a visitor in coastal areas from Pembrokeshire to the Wash. Any tiny, very active bird in trees or shrubs should be looked at closely.

GREAT TIT ■ *Parus major* Length 14cm; wingspan 22.5–25.5cm

DESCRIPTION This bird has a distinctive head pattern of black and white. It has a black centre stripe on the yellow breast and belly, a greenish back, blue-grey wings and tail, one white wing-bar and a white outer tail. In the male the black belly-stripe is noticeably wider between the legs; the female's is narrower. The juvenile's white face is washed yellow until the autumn moult.

HABITAT Found mainly in woodland, and also in parkland, orchards, hedgerows and town gardens.

HABITS This is a bold bird that is regular at birdfeeders and nest boxes. It feeds on a large variety of insects, spiders and larvae; seeds and fruits are taken in winter. The nest is constructed of moss lined with hair and feathers in a tree-hole, wall crevice or nestbox. The female lays 5–12 white eggs with red-brown blotches. Voice: the species has a complex, large vocabulary. The basic call is a sharp *chink*. The male usually has 3–4 distinct songs based on the typical *tea–cher*, although at a distance only one note may be heard; the phrase is usually sung at least half a dozen times in each song burst.

DISTRIBUTION This is the most widespread tit species in the world. It is found from the British Isles to Japan, and from Lapland in the north to Indonesia in the south. It is resident and sedentary in Britain. Although widespread, it is not as abundant in northern and eastern England, and Scotland (where it has spread northwards in recent years), as it is in the south. Shortly after breeding in territories, wandering winter flocks form – fewer birds visit garden feeders if there is a good beechmast crop in the woods than in poor crop years. Great Tits are tolerant of humans, and will feed from birdfeeders that are positioned very close to or even fixed to a window.

BLUE TIT ▪ *Cyanistes caeruleus* Length 11.5cm; wingspan 17.5–20cm

DESCRIPTION This is the region's only tit with a blue crown, outlined in white; there is also a dark line through the eye. The cheeks are white, outlined in black from the chin. The upperparts are yellowish-green, the underparts sulphur-yellow. Both the wings and the tail are dark blue, and there is one white wing-bar. The sexes are similar. Juveniles have a white face washed yellow, which is retained until the autumn moult.

HABITAT Chiefly a bird of broadleaved forest that has adapted to life with humans, and is now abundant in other habitats with trees, even including inner-city parks and gardens. It avoids conifers.

HABITS The Blue Tit wanders outside the breeding season, and is a favourite visitor to bird tables in winter. Naturally, it feeds high in trees, and displays inquisitiveness and agility in its quests to reach food. The young are fed on defoliating caterpillars. The Blue Tit readily uses nest boxes instead of natural holes in trees or walls, and lays 7–13 white eggs that are lightly speckled with red. Voice: the call is *tsee-tsee*, the alarm call *chirr.r.r* and the song a tremolo *tsee-tsee-tsee-tsuhuhuhu*.

DISTRIBUTION This species is widespread in summer and winter throughout Britain, except in highlands, Scottish islands and other areas with few trees. It is resident, so its summer and winter abundance are very similar; most birds are unlikely to move more than 10km. Populations erupt in years of high numbers. BTO Garden Birdwatch has reported that 98 per cent of Britain's gardens are visited by Blue Tits in winter, where their favourite foods are sunflower seeds and peanuts.

COAL TIT ■ *Periparus ater* Length 11.5cm; wingspan 17–21cm

DESCRIPTION The Coal Tit has a diagnostic glossy black cap and large white patch on the nape. It has white cheeks, a black chin, throat and upper breast, and buff underparts that are paler towards the centre. The upperparts, wings and tail are olive-grey, and it has two white wing-bars. The sexes are similar.

HABITAT This is the tit of coniferous woods, but also other woodland – it is found anywhere with firs, even in cemeteries, parks and gardens in cities.

HABITS Coal Tits feed on insects and spiders; they take seeds in autumn and winter, when they readily come to bird tables. Active birds, they are more agile than other tits when foraging high in the trees; they also search tree trunks, where they often cache seeds taken in the wild or from bird tables. They nest in a hole low down in a tree stump, wall or rock crevice, or even in the ground, and lay 5–11 white eggs with fine, reddish speckles. Voice: the call is a piping *tsee*, the song a loud, clear *teechu, teechu, teechu*.

DISTRIBUTION The species is found throughout the UK (except in the Scottish Highlands and Northern Isles, and the Fens in England). The summer and winter distributions are similar because it is relatively sedentary. It is generally less common than Blue and Great Tits (see p. 82 and p. 81). It has benefited from the planting of fir plantations, but is also common in Sessile Oak woods in the west, and in Silver Birch particularly in Scotland.

MARSH TIT ■ *Poecile palustris* Length 11.5cm; wingspan 18–19.5cm
Red listed

DESCRIPTION The Marsh Tit has a glossy black cap, a white face, and greyish-brown wings, tail and upperparts. The underparts are dull white, with a pale buff tinge on the flanks and under-tail coverts. The sexes and ages are alike. The species is separable from the Willow Tit (see right) with difficulty – note the Marsh Tit's glossy black cap, small black bib with well-defined edges, paler underparts, lack of a wing-patch, and quite distinct calls and song.

HABITAT Resident, spending all year in the same territory, preferably in extensive deciduous woodland that has a complex understorey where it most often feeds.

HABITS Despite its common name, this is not a marshland bird; its name first appeared in English in a 17th-century book as a translation of the original mid-16th-century Latin scientific name *palustris*, which means 'marshy'. Its food consists mostly of insects and spiders in spring and summer, and seeds and nuts in winter. It does not often feed at bird tables. The birds nest low down in natural holes, rarely in a nest box, and lay 7–9 white eggs with some reddish spots. Voice: the calls are *pitchoo* and a nasal *ter-char-char-char*; the song consists of the repetition of one note.

DISTRIBUTION Marsh Tits are widespread in England and Wales except in uplands. They are absent from Ireland and most of Scotland. Fragmentation of their preferred habitat has resulted in a considerable decline of nearly 70 per cent since the 1970s, which is still ongoing. Studies have revealed that the population is only one-tenth the Coal Tit's (see p. 84), and one-fiftieth that of the Blue Tit (see p. 82). Some people are fortunate to record this sedentary tit in their garden, if it wanders a little in autumn and winter around its small territory of 5–6ha.

WILLOW TIT ▪ *Poecile montanus* Length 11.5cm; wingspan 17–20.5cm

Red listed

DESCRIPTION The Willow Tit's cap extends to the mantle, and its bib is quite extensive with poorly defined borders. Note carefully the light patch on the secondaries of the closed wing, which the Marsh Tit (see left) does not have. Scandinavian and central European birds are greyer on the back and whiter on the face than British birds. Always observe the plumage and call notes to be sure of the identification.

HABITAT Found in mainland Europe in coniferous forest, trees in damp lowlands and mixed woodland. In Britain occurs regularly only in damp lowlands, especially alder carr, and woods close to rivers, lakes or reservoirs.

HABITS Feeds on invertebrates in the breeding season, and seeds and berries at other times. The species is noteworthy for the male and female excavating their own new nest-hole every year, low in a very soft, rotten tree stump. The female lays 6–9 white eggs with variable red markings. Voice: the calls are *eez-eez-eez* and a characteristic nasal *tchay, tchay*.

DISTRIBUTION The Willow Tit is a very sedentary resident, most abundantly in a crescent from north-east England to Wales; there are also more isolated populations in south-east Scotland, central southern England and Devon. It is often the most common tit in mainland northern Europe, but in Britain its population is well below that of all the other tits except the Crested Tit (see p. 88). Its 'lookalike', the Marsh Tit, has a population about three times as large, and the Coal Tit's (see pp. 84–85) is more than 15 times larger. The species has virtually disappeared from south-east England and the southern Midlands since the 1980s; the reasons for this are not clear, but are probably to do with the habitat becoming less suitable for it.

CRESTED TIT ■ *Lophophanes cristatus* Length 11.5cm; wingspan 17–20cm

Amber listed

DESCRIPTION An unmistakable bird, the Crested Tit is the size of a Blue Tit
(see pp. 82–83), and it is the only small bird with a crest, which is backwards-pointing
and black tipped with white. It has a distinctive face pattern of a curving black line on a
white face; there is a black line down the sides of the neck that joins the black bib. The
upperparts are buff-brown, the wings and tail grey-brown. The sexes are similar. Juveniles
have a shorter crest than adults.

HABITAT Occurs in pine forest in northern Britain, and in mixed or deciduous woods
elsewhere in Europe.

HABITS The Crested Tit is very sedentary, and is much less common than other tits. It is
a restless feeder as it seeks food, often hanging upside down, searching under branches. It
joins other tits and Goldcrests in winter as long as the group stays within its territory. Its
diet consists of insects, spiders and seeds, and it stores food in autumn for use in winter.
The species' range is restricted partly by its need for rotten wood in which to excavate its
nest-hole, but provision of nest boxes filled with wood shavings and sawdust can simulate
a dead stump and encourage a female to excavate a hole, in which a mossy nest is made
for the 4–8 white eggs with dark red spots. Voice: the vocal repertoire is limited to a low-
pitched, purring trill, and the song uses repeated calls.

DISTRIBUTION In Britain the 2,000 or so Crested Tit pairs are confined to northern
Scotland in pinewoods. They are most abundant in the pines, especially native Scots Pine,
from Moray and Nairn to the Caledonian pines in the Spey Valley. Adults are resident
throughout the year, but juveniles make short movements away from the home territory,
which gives Scottish garden birdwatchers a chance of seeing one, because they do visit
garden feeders.

LONG-TAILED TIT ■ *Aegithalos caudatus* Length 14cm; wingspan 14–19cm

DESCRIPTION This is a very attractive bird with the head and underparts whitish washed with pink, and a black stripe from the bill to the mantle. The upperparts, wings and tail are dull black; the scapulars and rump are pink, and there are white tips and edges to the 9cm-long tail. Juveniles are shorter and darker than adults, with little pink. The subspecies in northern and eastern mainland Europe has a pure white head and white-edged wings, and wanderers are sometimes recorded in the UK.

HABITAT Found in deciduous woodland, and thick scrub like gorse, bramble and briar.

HABITS Strictly speaking this is not a true tit as was once thought, and it is now in a small family of its own. Its flight appears to be weak, with the long tail noticeably trailing. The food consists especially of bugs, and insects' eggs and larvae. Territorial pairs take about three weeks to build an oval, domed nest of moss and spiders' webs covered with lichen; it has a side entrance near the top, and is lined with *c.* 1,000 feathers. Built in bramble or other thick cover, the structure's materials result in it being elastic, capable of expansion to cope with the growing 8–12 chicks. Non-breeding family flocks keep together through the winter with the help of *tsirrup* calls. These groups travel quite widely in winter, and increasingly in the last 25 years have been visiting gardens and feeding at bird tables. Voice: the song consists of a rapid repetition of calls.

DISTRIBUTION This species is a widespread resident throughout Britain, except in highlands and the Scottish islands. It is most abundant in summer and winter in England, roughly from Lancashire southwards. Due to its insectivorous diet it is very susceptible to cold winters, but it has large broods and numbers soon recover. Long-term population studies have revealed a nearly 100 per cent increase in numbers since the mid-1980s; the winter range in particular has expanded.

adult, northern race

SWALLOW ▪ *Hirundo rustica* Length 17–19cm; wingspan 32–35cm

Amber listed

DESCRIPTION The Swallow is familiar for its deeply forked tail with long, thin outer streamers (2–7cm long, with the male's being longer than the female's or juvenile's). The adult's upperparts and breast-band are shiny blue-black, the flight feathers dull black, the underparts off-white, and the forehead, chin and throat deep red. In juveniles the red is replaced by reddish-buff. Care should be taken not to confuse this species with the Swift (see p. 65), which is all black.

HABITAT Swallows occur in open country, in close association with humans; they are often found near water, engaging in aerial feeding.

HABITS The birds are gregarious on migration, gathering on wires and roosting, sometimes in their hundreds, in reedbeds. They feed on the wing on flying insects. Primeval nest sites are rare, in cave entrances. Nests are now mostly on ledges in farm buildings, sheds, garages, stables and church porches (see also House Martin, p. 92). The 4–5 eggs are white, lightly spotted with red-brown. Voice: the call is a sharp *witt* or *witt-witt*, and the bird has an attractive, warbling song on the wing.

DISTRIBUTION This species has the most extensive breeding distribution of Britain's summer visitors, from Shetland to the Isles of Scilly. It is least abundant in upland areas, especially in northern Scotland, and lowland arable areas in south-east England, and absent from central London. In recent years there has been an increase in the number of winter records, especially from coastal areas. Swallows winter in Africa, mostly south of the Equator.

adult at nest

HOUSE MARTIN ■ *Delichon urbica* Length 12.5cm; wingspan 26–29cm

Amber listed

DESCRIPTION The House Martin's upperparts are deep metallic blue; the flight feathers and tail are dull black, and sharply contrast with the pure white underparts and rump. The tail is deeply forked, the bill is small and black, the white legs are fully feathered and the feet are pink.

HABITAT The species is aerial over cliffs, villages and farms. It is a colonial nester, mostly in groups of five or less.

HABITS The House Martin's diet consists of flying insects. Few birds still nest on primeval sites under the rocky overhangs of cliffs; most now nest under the eaves of buildings. They build a cup-shaped mud nest lined with grass and feathers up against an overhang, with an entrance at the top; 3–5 white eggs are laid in the nest. Garden birdwatchers can hope to get breeding birds by erecting a batch of special nest boxes. Voice: utters a hard *chirrrp*, and twittering.

DISTRIBUTION House Martins are summer visitors. They breed in any place with suitable nest sites, except the highest parts of Scotland, the Outer Hebrides and Shetland. There has been an overall decrease in the number of birds of 14 per cent in England, but a 104 per cent increase in Scotland. The reason for this difference is unclear. House Martins winter across Africa south of the Sahara.

a cluster of birds ready to migrate

SKYLARK ■ *Alauda arvensis* Length 18–19cm; wingspan 30–36cm

Red listed

DESCRIPTION The Skylark is a streaky grey-brown bird that walks rather than hops. The streaks on the breast are sharply divided from the rest of the white underparts. In flight it shows a white trailing edge to the broad wings, and white outer-tail feathers.

HABITAT Found on farmland, heaths, moors, coastal dunes, and even airfields and golf courses. It rarely perches on trees.

HABITS The Skylark is most often noticed when the male flies up steeply, singing a very varied, melodious, non-stop outpouring on a 'song post' in the sky, *c.* 100m up, for several minutes, followed by a slow descent. The birds' diet comprises insects and seeds. The female lays 3–5 greyish eggs spotted with brown in a well-concealed nest on the ground, in short vegetation. Voice: utters a liquid *chirrup*, a distinctive flight note. Also see song above.

DISTRIBUTION Skylarks are widespread in most of Britain except at high altitudes. They are least abundant in built-up areas, the highest uplands and wetlands. More than two-thirds of Skylarks are found on eastern lowland farms, and nearly 40 per cent are located in cereal fields. There has been a widespread decline in their numbers of nearly 60 per cent since 1970 due to intense grassland management and the switch to sowing cereals in autumn rather than spring. Farmers are being encouraged and helped financially to leave patches of bare ground in large fields as nest sites – work that is being developed by the RSPB. The species is mostly resident; northern and eastern populations migrate to winter in the milder west. Near a Skylark habitat in calm weather, the bird's song can be heard from a garden, even if it is a kilometre or more away.

GARDEN WARBLER ■ *Sylvia borin* Length 14cm; wingspan 20–24.5cm

DESCRIPTION This warbler has a rounded head with a pale eye-ring, and a stubby bill. It has brown upperparts (darker on the wings), and is pale buff below, fading to white on the belly and under-tail coverts, so the plumage is plain above and below. The adults of both sexes are alike. More often than not this species is identified by its song (see below).

HABITAT Breeds in broadleaved woodland with thick undergrowth, scrub and young conifers.

HABITS The Garden Warbler avoids competition with the Blackcap (see pp. 96–97), its closest relative, by arriving later than that species and feeding lower down. Its food consists mostly of insects, plus some fruits including berries. The 4–5 whitish eggs with variable markings are laid in a nest partly made by the male and finished by the female, in low vegetation. Voice: the call is a loud *tac tac*; the song is a rich, sustained warble that can be confused with the Blackcap's song but is pitched lower, with longer phrases, and sometimes sung for up to a minute from thick cover, so that the bird is hard to see.

DISTRIBUTION The Garden Warbler is a summer visitor and is widely distributed in ideal habitat from southern and central Scotland southwards. It is most common in well-wooded parts of western Britain, and least abundant in intensively farmed land. It is very scarce in Northern Ireland. Western European populations winter in western Africa, while eastern populations do so in eastern and southern Africa. Despite its name this species is rare in gardens. It is most likely to occur – as a singing male passing on spring migration – in gardens that replicate its natural habitats, or are close to them.

GRASSHOPPER WARBLER ■ *Locustella naevia* Length 12.5cm; wingspan 15–19cm

Red listed

DESCRIPTION This warbler's upperparts are olive-brown, spotted and streaked with dark brown. Its wings are darker with buff to reddish fringes, and the tail is reddish-brown, barred darker. The underparts are mostly buff, but almost white on the chin, throat and centre of the breast and belly. The under-tail coverts are streaked with brown. The sexes are alike.

HABITAT Occurs in marshes, thick hedges, heathland and new plantations.

HABITS This skulking bird is very difficult to see. It eats mostly insects in early summer, and also berries and other fruits before its autumn migration. The nest is constructed on or near the ground in a tussock of grass or sedge, hidden by overhanging stems. The female lays 5–6 pinkish eggs heavily marked with red-brown speckles and dark lines. Voice: the species' strange song, voiced mainly at dusk or dawn, is uniformly pitched and sustained, like that of trilling insects, and far carrying, even up to 1km, but it is inaudible to some people. The bird sings seemingly without pause, but actually with very short breaks. One so-called 'uninterrupted' song lasted 110 minutes.

DISTRIBUTION This summer migrant is widely but thinly distributed throughout Britain. It breeds in suitable habitat most commonly only in Wales, western Scotland and central England. It is less common than it was years ago, most probably because of loss of habitat. This can be one of the rarest and strangest sounding birds in a garden that borders ideal habitat; most likely it will occur on just one day in the form of a spring migrant still on passage.

BLACKCAP ▪ *Sylvia atricapilla* Length 13cm; wingspan 20–23cm

DESCRIPTION This species is easily confused with Marsh and Willow Tits
(see pp. 86 and 87). The adult male has a black crown, and an ash-grey nape and face.
The upperparts are ashy-brown, and the tail and primaries are darker. The chin, breast
and flanks are grey, and the belly and under-tail coverts are white. The adult female is
similar but her cap is bright red-brown and her upperparts are browner; immatures look
very like females.

HABITAT Occurs in open woodland, copses with thick undergrowth and parks.

HABITS Blackcaps forage and sing in tree-tops. Their diet consists mostly of insects in
summer and, especially in autumn, fruits including berries. In winter they are very fond of
feeding on fat on the bird table or berries on the garden's bushes. The female lays 4–6 eggs
with a very variable ground colour and dark spots, in a neat nest low in a bush, creeper or
brambles. Voice: the call is a loud, repeated *tac*; the song is a loud, rich warble, rising in
pitch, not as sustained as the Garden Warbler's (see p. 94).

DISTRIBUTION The Blackcap is usually thought of as a summer visitor that winters in
the western Mediterranean region. Increasingly in recent years, birds from central Europe
have been wintering in the British Isles, which is an example of a rapid evolutionary
change. The species is widely distributed in the UK except in highland regions in both
summer and winter, but is much more abundant in summer; in the winter it is most
common in the south, south-west and Welsh Borders. Some birds winter in the British Isles
in parks and gardens in town, and the species is the warbler most often recorded in gardens.

♀

WILLOW WARBLER ▪ *Phylloscopus trochilus* Length 10.5–11.5cm; wingspan 17–22cm

Amber listed

DESCRIPTION Olive-green above and yellowish-white below, Willow Warblers have cleaner colours than the similar Common Chiffchaff (see pp. 100–101). The supercilium is pale yellow, and the legs are orange-brown. Adults in summer become browner above and whiter below from feather abrasion. Juveniles in autumn have a much yellower supercilium, throat and breast. *Phylloscopus* warblers are popularly known as leaf warblers due to their colours and feeding habitats.

HABITAT The Willow Warbler is a bird of mature woods, especially Sessile Oak and Silver Birch, favouring coppices and scrub.

HABITS This is an active little bird that searches the canopy for insects. It is difficult to see in the canopy, but often sings from a prominent perch. The well-concealed, domed nest is constructed on the ground, in thick herbage. The female lays 4–8 white eggs marked with reddish spots. Voice: the call is a plaintive disyllabic *hoo-eet*; the song is a lovely cascade of pure notes, lasting for three seconds and dying away towards the end.

DISTRIBUTION This species is a summer visitor, and is the most common of the leaf warblers in Europe, although it has declined by nearly 40 per cent in recent years due to a complex set of factors. It winters in tropical Africa beyond the Sahara. Because of its habitat preference it is mainly found in western Britain where its favoured trees are found. It is particularly abundant in the wooded uplands of south-west England, Wales, Cumbria and the western Highlands of Scotland.

juvenile (above), adult (below)

COMMON CHIFFCHAFF ▪ *Phylloscopus collybita* Length 10–11cm; wingspan 15–21cm

DESCRIPTION This species is like a less streamlined Willow Warbler (see pp. 98–99), with a less yellow tint. It is dull brownish-olive above and dull pale yellow below, shading to buff flanks. Eastern forms are noticeably greyer above and whiter below. The birds have a pale yellow supercilium and pale eye-ring, and a contrasting dark eye and dark legs (compare with Willow Warbler; it also appears duller plumaged and dumpier).

HABITAT The Common Chiffchaff is mostly thought of as an early-arriving summer visitor to open woodland with a good shrub layer, copses and hedgerows.

HABITS The birds are active insect hunters. They build a domed nest on or near the ground in thick cover such as long grass, brambles, gorse or nettles; the female lays 4–7 white eggs that are lightly spotted reddish. Voice: the call is a monosyllabic *hweet, hweet;* the song is a diagnostic *chiff-chaff-chiff-chaff…*, with more variation than the name suggests. This is one of the few birds that says its name.

DISTRIBUTION This is the second most common leaf warbler in Britain; it is a summer visitor, breeding throughout the region except in highland areas where a shrub layer is lacking. Many rural and suburban gardens have a singing bird in spring, and maybe in autumn. Most Chiffchaffs winter around the Mediterranean and across Africa south of the Sahara, but increasingly in the last 30 years 500–1,000 birds have been wintering in the British Isles, mostly in the south and by the coast, in England in particular. In these cases it is found especially in areas with water such as water-treatment works, and ponds and streams with thick cover where insects remain; it remains insectivorous, so cold winters kill many birds.

WAXWING ■ *Bombycilla garrulus* Length 18cm; wingspan 32–35cm

DESCRIPTION The Waxwing is an unmistakable bird with a noticeable crest sweeping back from its crown. The head, back and wing-coverts are vinaceous brown, the rump is grey and the tail is black with a yellow tip. There is a narrow black eye-stripe with chestnut above and below, a black bib and a reddish vent. The wing pattern is diagnostic: black or blackish flight feathers, white-tipped primary coverts and white-tipped secondaries with wax-red tips beyond, and tips of primaries edged with white and yellow.

HABITAT Occurs in lowland and upland conifer forests, where it breeds in Scandinavia

and Finland. In Britain it can be seen anywhere in the right season (see below), on the east coast especially, where it can find berries.

HABITS In the breeding season and in warm winter weather, Waxwings feed by catching flies from tree-tops. Their food consists mainly of insects in summer, and berries in winter. Voice: the call is a ringing *sirrrrrr*; the birds have a twittering song.

DISTRIBUTION Waxwings are seen in Britain only in autumn and winter, and there are many more in some years than in others. This species is famous for its irruptions as dozens, hundreds or even thousands of birds cross the North Sea to Britain in some winters from northern and eastern Europe because the berry crop there has failed. When this happens, large winter flocks irrupt to find trees and bushes with red berries as far west as Britain, France and Ireland. At these times they are a marvellous sight in gardens, feeding on berries of plants such as cotoneaster and rowan. Waxwings are most likely to be seen in gardens near the east coast; comparatively few are seen in south-west England, Wales Northern Ireland and north-west Scotland.

1st winter

NUTHATCH ■ *Sitta europaea* Length 14cm; wingspan 22.5–27cm

DESCRIPTION The Nuthatch is like a small, blue-grey woodpecker, with a strong black bill, short tail and rounded wings. The upperparts are blue-grey, the cheeks and throat are white, and the rest of the underparts are orange-buff. There is a broad black eye-stripe, and the outer-tail feathers are black with white subterminal spots. The sexes are similar. Juveniles are duller below than adults.

HABITAT Prefers broadleaved woods, but also occurs in mixed woods and gardens with mature trees. A pair lives in the same territory throughout the year.

HABITS The diet consists of invertebrates and seeds, and Nuthatches regularly visit bird tables. They are named from their habit of fixing a seed in a crack and hammering it with the bill. The birds may be seen at feeders, picking up seeds and repeatedly flying to nearby trees, where they jam the seeds in a crack for future reference. The Nuthatch is Britain's only bird that can move head-first down a tree trunk; even woodpeckers cannot do this. The female builds the nest in a tree-hole or box, plastering the nest-hole with mud to reduce its size to ensure that only she can get into the nest, and not a larger predator. She lays 6–8 white eggs heavily marked with reddish-brown and grey. Voice: the call is a loud *chwit-chwit*; the song consists of a rapid *chu-chu-chu-* and a slow *pee, pee, pee*.

DISTRIBUTION The Nuthatch is sedentary, so its summer and winter distributions are the same. It is widespread in southern Scotland and England (except the Fens and south Yorkshire), and most common in Wales and southern England. It is not found in Ireland. It was not proved to breed in Scotland until 1989, since when it has continued to steadily increase its range.

TREECREEPER ▪ *Certhia familiaris* Length 12.5cm; wingspan 17.5–21cm

DESCRIPTION The Treecreeper is brown above and white below, relieved by a rufous rump, white supercilium, mottled and streaked back, two pale wing-bars, distinctive buff band across the wing and white-spotted tertials. The tail is long and brown, with stiff, pointed feathers; the bill is quite long and gently decurved.

HABITAT Found in coniferous forest in mainland Europe, but in deciduous forests in the British Isles.

HABITS This bird is hard to observe because of its cryptic colours and high-pitched voice. It has the very distinctive feeding behaviour of searching for insects and spiders in a crouching walk *up* one tree, and flying to *low down* on another. In some places its movement on tree-trunks has earned it the old name of Tree Mouse. It usually nests in a gap between tree bark and the tree to which it is attached. The female lays 1–6 white eggs with pink or reddish-brown spots. The call is a thin *tsiew*; the song is a cadence lasting 2.5–3 seconds.

DISTRIBUTION Widely distributed throughout the British Isles except the Northern Isles, highlands, and Fens. The species is resident and sedentary. Numbers are hard hit by prolonged frosts. It may occur in gardens with mature trees and may even be attracted to a specialist nest box (see www.nestbox.co.uk) – though Blue Tits may well take this over.

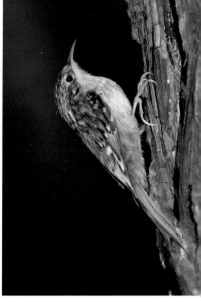

WREN ▪ *Troglodytes troglodytes* Length 9–10cm; wingspan 13–17cm

DESCRIPTION The Wren is a tiny, unmistakable bird that weighs only 7–12g. It is reddish-brown above, whitish-brown below, with fine, darker barring, and has a white supercilium. It has the habit of holding its short tail vertically. The bill is pointed and quite long. The sexes and ages are alike.

HABITAT Found in a wide variety of habitats with dense undergrowth: woodland, islands, gardens, hedgerows, reedbeds, and moorland with heather and Bracken.

HABITS Wrens employ a rapid, straight, whirring flight between bushes. They are active, hunting invertebrates, insects and spiders in low cover. The male builds several completely domed nests; the female chooses one and lines it with feathers. She lays 5–8 white eggs, some with dark speckles at the broad end. Voice: utters a rattling alarm call, and a hard *tic tic tic*. The song consists of notes and trills uttered for 4–6 seconds; it is amazingly loud for a small bird, and is sometimes sung from an exposed perch. Although this does provide an opportunity to see the bird, it is more often heard than seen. Males hold territory all year round, so the song can be heard in every month.

DISTRIBUTION The Wren is one of Britain's most widespread species, found in a wide variety of habitats. Its preferred habitat is woodland, but when the population is high Wrens move into other areas such as parks and gardens. The species is mainly resident; it suffers in hard weather when invertebrate food is hard to find.

STARLING ■ *Sturnus vulgaris* Length 21.5cm; wingspan 37–42cm

Red listed

DESCRIPTION This is a dumpy, short-tailed bird, smaller than a Blackbird (see pp. 112–113). Its black summer plumage has a green-and-purple sheen, and it walks with a waddle, not the Blackbird's hop. The winter plumage is covered with whitish spots. In winter the pointed bill is dark, and it turns bright yellow in the breeding season. Juveniles are a dull, dirty brown with a whitish throat; they moult in the autumn to the adult winter plumage, with the head being the last to change. The birds have a delta-winged shape in flight.

HABITAT Occurs on farmland and in suburbs, especially where there is short grass including lawns.

HABITS Winter flocks and roosts are preceded by spectacular, swirling flights (murmurations), with the birds sometimes numbering in the thousands. Starlings feed on short grass, digging with the pointed bill for invertebrates, especially cranefly larvae, or leatherjackets; they also eat caterpillars, seeds and fruits. They nest in holes in buildings, and also in trees. The female lays 4–6 plain blue eggs. Voice: the call is unusual, a grating *cherrr*; the song is a prolonged series of whistles, squeaks, warbles and mimicry (the latter including the sounds of curlews, doves, farmyard birds, ringing telephones or any local, repeated sound).

DISTRIBUTION The Starling is a widespread resident in summer and winter, except in the Scottish Highlands. It is most abundant in low-lying, pastoral farmland, and least common in Wales, south-west England and the Scottish uplands. There has been a 50 per cent decline in the species in Britain since the 1970s, especially where it is now least common. The decline is believed to be due mainly to poor survival of first-year birds. Their preferred habitat, pastureland, is now so managed that it often has a poor supply of the favourite food, cranefly larvae. Young birds in particular disperse and roam fields in flocks. Starling numbers are greatly swollen in winter with the arrival of thousands of visitors from northern and central Europe. So your garden visitors may be foreigners, such as Poles or Germans.

winter adult *summer adult*

MISTLE THRUSH ▪ *Turdus viscivorus* Length 27cm; wingspan 42–47.5cm

Amber listed

DESCRIPTION In adults the whitish underparts are covered with large, wedge-shaped black spots; the flanks and breast are marked with buff. The upperparts and wings are greyish-brown with conspicuous greyish-white fringes. The tail is grey-brown with diagnostic white tips to the outer feathers, which can be clearly seen when the bird is flying away from you. The sexes are similar. The juvenile's plumage is noticeably different from that of the adults – it is spotted white on the head, mantle and wing-coverts.

HABITAT Found in orchards, woods, farmland, parks and large gardens.

HABITS The white underwing and tail marking is striking in flight. The diet comprises invertebrates, and in autumn and winter the birds eat berries, especially those of holly (they also eat mistletoe if this is available). Where there are mature deciduous trees in a garden or near it, they may be the territory of a pair of Mistle Thrushes. The territory is vigorously defended, and the birds nest in one of the trees, not in a bush like their relatives. The female lays 3–5 creamy-buff eggs with red-brown, grey and lilac markings. Voice: the call is a harsh, distinctive rattle. The song is short and loud, consisting of flutey phrases, which are far carrying from a high vantage point in a tree, even in wild weather – hence the bird's country name of Storm Cock.

DISTRIBUTION Mistle Thrushes are largely sedentary, breeding and wintering throughout the UK except the Outer Hebrides, Northern Isles and highest parts of Scotland. Populations move south and west in autumn and winter into milder parts of the breeding range, forming small flocks. The species is nowhere as abundant as its close relatives, with less than half a million pairs, compared with two million Song Thrushes and more than four million pairs of Blackbirds.

SONG THRUSH ■ *Turdus philomelos* Length 23cm; wingspan 33–36cm

Red listed

DESCRIPTION Song Thrushes are brown above, and white below with a buff tint to the breast and flanks; the underparts are streaked with black-brown, arrow-shaped marks (not spotted like those of the Mistle Thrush, see pp. 108-109). There is an orange-buff underwing, which is visible in flight.

HABITAT Occurs in woodland, parks and gardens with plenty of shrubs.

HABITS Feeds on invertebrates, fruits and snails; the last are eaten mostly in late summer and through the winter, and earthworms and caterpillars are the main foods in spring and early summer. Snails' shells are broken by being beaten on a stone – the litter of shells by a stone is a clear sign of a 'thrush's anvil'. The nest of grass and moss is uniquely lined with mud and decayed wood; the female lays 3–5 distinctive blue eggs that are sparsely spotted with black. Voice: the call is *sipp*. The alarm call is a rapid, repeated, scolding *tchuk-tchuk-tchuk*. The song is a loud, sustained, characteristic series of phrases, each often repeated 2–4 times.

DISTRIBUTION Most birds are resident and breed throughout Britain, except in the Scottish Highlands. The species is most abundant in summer and winter in lowland areas in the south and west. Since the 1970s there has been a population decline of more than 50 per cent, but there are signs of recovery. In the early 20th century it was more common than the Blackbird; now the reverse is true.

BLACKBIRD ■ *Turdus merula* Length 24–25cm; wingspan 34–38.5cm

DESCRIPTION The adult male is the only all-black European bird with a bright golden-yellow bill and long tail, and it has an orange-yellow eye-ring. In the immature male the bill is dark brownish-horn, and the plumage is dull black. The female is a 'brown bird' – all dark brown, often with a rufous tone to the underparts and a thrush-like mottling on the breast. Juveniles of both sexes before the autumn moult are like the female but more rufous, and more spotted below. The adult plumage appears during the autumn moult, which creates some unusual-looking males with black plumage, except for the last part to change – the brown head. Some people confuse the male with a Starling (see p. 107), which is dumpy, shorter tailed and has a purple-and-green gloss in summer, and in winter is covered with white spots. Not infrequently Blackbirds are reported with partly white plumage – a genetic fault – and such plumage may become whiter in ensuing years. These 'white' birds still have dark eyes so are not albinos which have pink eyes and pale beak and legs. This colour form is scientifically called 'leucistic'.

HABITAT Breeds in most places where trees are present, but also on moors.

HABITS The Blackbird was originally a forest bird, but now 30 per cent of the population is in urban parks and gardens, where life is nearly twice as safe for it as it is in woodland despite the threat posed by domestic cats. It feeds on insects, earthworms and, in autumn and winter, wild fruits. It hunts in leaf litter with a distinctive action, flicking its bill to and fro, and disturbing the leaves to reveal food. Two or even three broods are reared in a year, in a deep cup nest in a bush, creeper, crevice in a bank or brambles. The female lays 3–5 greenish-blue eggs freckled with brown. Voice: the call is *see*, the alarm call is a shrill chatter and the song consists of a variety of flute-like, bold musical phrases.

DISTRIBUTION The Blackbird is one of Britain's most widespread and common birds. It is resident, breeding and wintering everywhere except the highest uplands. It is most abundant in lowland England, and least common in the uplands of England, Wales and Scotland. In winter numbers are swollen by immigrants from mainland Europe. In spring a British male can be distinguished from a foreign one by its yellow eye-ring and bill; a migrant's bill will still be black.

male flight

male singing

♀

FIELDFARE ▪ *Turdus pilaris* Length 25.5cm; wingspan 39–42cm

Red listed

DESCRIPTION This unmistakable large, beautifully plumaged thrush has a slate-grey head, nape and rump that contrast with the chestnut back, black tail and white underwings. The throat and breast are golden-brown streaked with black. The rest of the underparts are white, and the flanks are streaked with black.

HABITAT Breeds in open woodland, scrub, gardens and parks.

HABITS The Fieldfare is a gregarious species, and many hundreds of birds roost together in evergreens in winter; it most commonly breeds on the Continent in colonies of up to 40–50 pairs. It feeds on a variety of invertebrates, and fruits in autumn and winter, and all feeding is done on neutral ground. It nests in a bush or tree, and the female lays 4–6 eggs very like a Blackbird's. The Fieldfare is a noisy, aggressive bird at nest sites and when defending winter food sources such as bushes with berries. Voice: the call, a loud *tchak tchak*, readily identifies migrating flocks. The song is a weak warble with some wheezes and chuckles.

DISTRIBUTION This species is primarily a winter visitor to Britain from Fennoscandia. Visiting birds are found throughout the UK except the Scottish Highlands and large cities. They are most abundant in low-lying land in England where there is suitable food on open ground, and in fruit-laden bushes and trees. A very few pairs nest in Britain each year.

REDWING ■ *Turdus iliacus* Length 21cm; wingspan 33–34.5cm

Red listed

DESCRIPTION All of the Redwing's upperparts are dark warm brown, darker on the flight feathers. The breast is yellowish-buff on the sides, with dark brown streaks. The underparts are white streaked with lighter brown, and the flanks and underwing are noticeably chestnut-red, hence the species' name. The bird may be confused with the Song Thrush (see pp. 110–111), which has buff underwings; also note the Redwing's white stripe above the eye and another below its dark brown cheeks. The sexes are alike.

HABITAT Nests in open woods, thickets and scrub, but winters on grassland and arable fields, and in open woodland.

HABITS Night migrants' calls keep a flock together and alert observers below. In winter the birds roost in large flocks in thick cover. The diet consists of a wide variety of invertebrates, and berries in autumn and winter. The birds build a cup-shaped nest in a tree or bush, or even on the ground. The female lays 4–5 bluish eggs heavily speckled with reddish-brown. Voice: the flight call is *see-ip*; the song in the breeding season consists of 4–6 flutey notes plus warbling, which is sometimes heard from birds in a flock in spring preparing to migrate.

DISTRIBUTION The Redwing is primarily a winter visitor to the UK, arriving from late September onwards, mostly as a night migrant. It breeds across Fennoscandia, into Siberia (from whence come most of Britain's visitors), and in Iceland (these birds winter in Ireland and western Scotland). There is a small annual breeding population in northern Scotland of fewer than 50 pairs compared with about a million in Sweden. The species is very nomadic when looking for food, and comes into gardens for berries when cold weather drives it off farmland.

ROBIN ■ *Erithacus rubecula* Length 14cm; wingspan 20–22cm

DESCRIPTION The upperparts, wings and tail are olive-brown, and the forehead, cheeks and breast are orange-red, separated from the brown by a band of blue-grey. The flanks are warm buff, and the rest of the underparts are white. The sexes are alike in adults. Juveniles are quite different. They are brown above, and buff below (with no orange), with buff spots all over the brown upperparts, and small, darker spots on the breast.

HABITAT A regular in parks, gardens, farmland, woods and copses with open spaces on which to feed.

HABITS Robins are often aggressive at bird tables, driving off other species. Their diet consists of insects, especially beetles; they will follow animals including humans to seek disturbed food, for instance when a garden is dug. This is an instinct from ancient times, when Robins fed on invertebrates disturbed by moles, deer, wild cattle and pigs in forests. Robins naturally nest in holes and crevices in the ground, and will use open-fronted nest boxes. Each year, a pair may produce up to three clutches of 3–7 white or pale blue eggs with reddish spots. Voice: the call is a scolding *tic tic* and high-pitched *tswee*. The song is a melodious warble, sounding sadder in autumn than at other times. Males defend a territory all the year round, so the song can be heard in every month. In distinctive display behaviour a Robin points its bill upwards and shows off its orange-red breast. Many females defend a winter territory, too. Robins can become hand tame if they are provided with mealworms.

DISTRIBUTION The Robin is one of Britain's most widespread and well-known birds. It is found throughout the year; most UK birds are resident, but winter numbers are swollen by migrants from continental Europe. The species is most abundant from central England southwards, and becomes much less common northwards and at altitudes above 400m.

juvenile

NIGHTINGALE ■ *Luscinia megarhynchos* Length 17cm; wingspan 23–26cm

Amber listed

DESCRIPTION The Nightingale is uniform brown above with a plain rusty-red tail and rump, and pale greyish-brown underparts with a paler throat and belly. There is a noticeable black eye with a whitish eye-ring. The sexes are alike.

HABITAT Mostly solitary, with traditional territories in thickets and woods with rich undergrowth, near water; it is especially fond of coppiced woods.

HABITS Nightingales feed on invertebrates and nest on the ground. The female lays one clutch annually of 4–5 glossy olive-brown eggs. The birds are skulking and hard to see, but much loved in literature and folklore because of their splendid song, a very varied, melodious succession of phrases, whistles and trills, with a recurring group of rapid *chooc chooc chooc* notes, and higher-pitched, flute-like, slow *pioo pioo pioo* in a rising crescendo.

DISTRIBUTION Sadly this summer visitor and famous songster is confined to southern and south-east England, and has declined by 90 per cent in the past 40 years. It is found roughly only south of a line drawn from the Wash to Dorset, and there only patchily in its ideal habitat. BTO surveys in 2012/2013 found that there were just over 3,000 territories in England. The Nightingale is most likely to be heard or seen in a garden in Kent, Sussex or Essex. Its decline is due to a great deterioration in suitable woodland because of less management, and over-grazing by deer.

COMMON REDSTART ■ *Phoenicurus phoenicurus* Length 14cm; wingspan 16–23cm

Amber listed

DESCRIPTION All ages and both sexes have an orange-chestnut rump and tail (with a dark centre). The adult male has a striking head pattern, with a white forehead, grey crown, and black face and throat; the rest of the upperparts are blue-grey, including the wing-coverts; the wings are blackish-brown, and the upper breast is black and sharply divided from the orange underparts. The adult female and juveniles have grey-brown upperparts and pale orange-buff underparts, fading to a whitish belly.

HABITAT Breeds mostly in old deciduous woodland and hill country with scattered trees and stone walls.

HABITS The tail is very noticeable in flight and when perched, quivering up and down. The diet consists mainly of insects, and some fruits. The birds nest in a hole in a tree or wall, and the female lays 5–7 blue eggs. This species may be tempted to nest in a garden from nearby woods if nest boxes are provided. Voice: the call is a plaintive *wheet*, often linked with a *tooick*, especially when the bird is alarmed near its nest; the song is brief and melodious.

DISTRIBUTION This is a summer visitor to Britain except Northern Ireland, particularly in northern and south-west England, Wales and Scotland in suitable habitat. Its distribution has decreased by more than 30 per cent since the 1970s. It winters in tropical Africa north of the Equator. If a garden in Britain borders ideal habitat the gorgeous male may be seen, especially in spring when he sings from a prominent, often high perch.

1st autumn

BLACK REDSTART ■ *Phoenicurus ochruros* Length 15cm; wingspan 23–26cm

Amber listed

DESCRIPTION All ages and both sexes have a dark-centred orange tail like that of a Common Redstart (see pp. 118-119). The adult male is dark grey above with dark underparts, blackest on the face and greyest on the belly; there is orange only on and under the tail. The wings are brownish-black with an off-white panel. The female and immatures are mouse-grey, and the tail-end colour is less bright.

HABITAT In mainland Europe Black Redstarts are found on stony, craggy hillsides, regularly by farmsteads and in villages, but in Britain most nest in industrial or city sites on large buildings.

HABITS The species often winters on rocky coasts, and in recent years has even done so in towns. It feeds on invertebrates and fruits. The female lays 4–6 white or sometimes pale blue eggs in a nest in a hole or crevice in a cliff or wall. Voice: the call is a quiet *tsip*, often preceding the alarm call, *tuc tuc*. The song is a quick, quiet warble; its strange ending sounds like crackling cellophane.

DISTRIBUTION The Black Redstart is a summer visitor to south-east England; at least 20 territories are in Greater London out of only 50 or so nationally. It is mostly seen as a regular autumn and winter visitor, arriving from France eastwards, to all parts. The birds are most common on or near the coasts of England and Wales, especially in the south and south-west, and they wander into coastal villages and towns.

adult male

1st winter

adult male

STONECHAT ▪ *Saxicola torquata* Length 12.5cm; wingspan 18–21cm

DESCRIPTION The adult male is distinctive, wih a striking white collar to its black head. Its back is dark brown, and its white rump is streaked brown. The wings are dark brown with a variable-sized white panel. The breast is orange, shading to white on the belly. The adult female has a mottled brown head and upperparts, and less bright underparts. The juvenile is like a heavily spotted female.

HABITAT Mostly resident in gorse and other scrub, heaths, sand-dunes and young plantations; shuns intensive agriculture.

HABITS The Stonechat is a restless bird, constantly flicking its wings. Its diet consists mainly of insects and spiders, which are mostly caught on the ground having been spotted from a perch. It nests on or near the ground in long grass under a bush. The female lays 4–6 pale blue eggs with some reddish markings. Populations suffer in cold weather when the species' insect food is hard to find. Voice: the call is *tchak*, sounding like stones being knocked together; the song consists of short, scratchy phrases.

DISTRIBUTION This species is mostly found breeding in southern, western and northern Britain. It is very scarce in the Midlands, and the east and south-east. It is recorded more widely in winter, a time when it may wander into gardens, but its greatest abundance still mirrors the breeding distribution. About half of British birds are resident, and their population suffers in hard winters. The other half migrate to the Mediterranean basin.

1st winter

SPOTTED FLYCATCHER ■ *Muscicapa striata* Length 14.5cm; wingspan 23–25.5cm

Red listed

DESCRIPTION The adults' upperparts, wings and tail are plain grey-brown. The forehead and crown are streaked black outlined in white. The underparts are white, washed with brown on the sides of the breast and flanks. The juvenile is *the* spotted flycatcher: its upperparts are buffer than those of adults; the head, back and wings have pale, round, buff-white spots; the underparts are not streaked but spotted dark brown.

HABITAT Often the last summer visitor to arrive in woodland edges, parks, orchards and gardens.

HABITS The Spotted Flycatcher obtains most of its insect food in flight, sallying forth from a perch, then often returning to the same perch. The nest, in which 4–6 bluish eggs with red blotches are laid, is formed on a ledge or in a tree fork, creeper or open-fronted nest box. The species is often tolerant of humans, even nesting very close to houses. Voice: the call is *tzee-zuk-zuk*. The song is quiet and short, and sounds like a squeaky wheelbarrow.

DISTRIBUTION This is a summer visitor throughout Britain except the Scottish islands. It winters in Africa, mostly south of the Equator, which is where it is thought that the problems are that have resulted in the birds' current scarcity – the population has declined by nearly 80 per cent since the 1960s to barely a quarter of what it used to be. It is most likely to occur as a garden visitor for maybe an hour or so in autumn, at the beginning of its migration south.

1st winter

DUNNOCK ■ *Prunella modularis* Length 15cm; wingspan 19–21cm

Amber listed

DESCRIPTION Warm brown above, and with lead-grey head and underparts, the Dunnock recalls a female House Sparrow (see pp. 126–127). Before the 1950s its English name was Hedge Sparrow, but the two species are in completely different families; note this bird's pointed bill and the House Sparrow's stubby, seed-eating bill. The brown upperparts are streaked with blackish-brown, the sides of the breast and flanks are buff-brown, streaked darker, and the crown and ear-coverts are brownish.

HABITAT Occurs in woodland along river valleys, hedgerows, spinneys, gardens and low scrub on moorland.

HABITS Dunnocks happily nest in gardens with plenty of bushes. They are unobtrusive ground feeders, creeping along, belly near the ground, looking for insects and seeds. By contrast, they often sing from exposed perches. They most often feed under feeders on what finches and tits spill, but do learn to fly up to a table or birdfeeder. The female builds a nest of twigs lined with grass, moss and hair, in which she lays 4–6 beautiful plain blue eggs. Voice: the call is a loud *tseep*, the song a short, loud, high-pitched warble. A trio of birds may be seen in a garden in spring, rather than a pair; the birds indulge in mad chases around the garden and lots of wing flicking when they perch. The female is attached to one particular male, but another male is constantly in attendance looking for a chance to mate with her; he also helps to feed the young.

DISTRIBUTION This is a very sedentary species in the British Isles, or it disperses over only short distances. It breeds and winters throughout Britain except in treeless areas in Scotland. It is most abundant in woodland, and gardens that are rich in trees and shrubs.

TREE SPARROW ■ *Passer montanus* Length 14cm; wingspan 20–22cm

Red listed

DESCRIPTION This species is similar to the House Sparrow (see pp. 126–127) but neater looking. The distinctive head pattern distinguishes it from the House Sparrow: there is a red-brown cap, a small black bib with a sharp bottom edge, white cheeks and incomplete collar, and a black patch below and behind the eye. It has double white wing-bars, with the lower one being less noticeable than the upper one. The sexes are alike.

HABITAT Occurs in open country with mature trees, orchards, old hedgerows and pollarded willows by slow-flowing rivers, and locally in gardens. Winters on stubble, in farmyards, and can become regular at feeding stations in reserves and gardens.

HABITS This species is less associated with humans in Europe than it is in Asia. The diet consists of seeds of grasses and cereals, and small invertebrates and their larvae. The birds nest in holes in trees and buildings, and in nest boxes (they will use the same site year after year). The female lays 2–7 off-white eggs that are variably marked with grey and brown. Voice: the call is more high pitched than the House Sparrow's, a sharp, repeated *teck*; the song is just a repetitive *chip*.

DISTRIBUTION The Tree Sparrow is a summer and winter resident. It is mostly confined to low-lying land in England north of a line from Bristol to Norwich, but only eastern Wales, then northwards to eastern Scotland and south-east Northern Ireland. The House Sparrow is about 20 times more abundant than this species. The Tree Sparrow is mainly sedentary, but northern populations sometimes irrupt. It has suffered a severe decline in the last few decades, much of it believed to be due to changes in agricultural practices; a recent increase still leaves the species numbering only 10 per cent of its population in the 1960s.

HOUSE SPARROW ■ *Passer domesticus* Length 14–15cm; wingspan 21–25.5cm

Red listed

DESCRIPTION In the male most of the upperparts and wings are chestnut streaked with black, the crown and nape are grey, the bib is big and black with a broken bottom edge, the cheeks are dirty white, the rump is grey and the tail is dark brown; there is one white wing-bar and the underparts below the black bib are dull grey. The female, not to be confused with the Dunnock (see p. 124), is brown above, with underparts that are all dull grey. The crown is olive-brown and the supercilium is pale buff, especially noticeable behind the eye.

HABITAT Breeds close to humans on farmland, and in towns and cities.

HABITS The House Sparrow is a ground feeder on wild seeds, cereal stubble, insects and their larvae (fed to the young), 'bird food' and scraps at bird tables. The nest is an untidy domed affair of grass and straw, and the birds breed in small colonies in holes in trees, buildings, nest boxes and House Martin nests. The female lays four off-white eggs that are variably speckled or blotched with grey and brown. The birds are sociable, noisy, chirruping and feed in flocks, although large post-breeding flocks are rare now. The song does not really deserve that title – it is a monotonous, repeated *chirp*.

DISTRIBUTION House Sparrows are sedentary, never wandering more than a few kilometres from their birthplace. This is one of Britain's most widespread birds, common everywhere except the uplands, especially those of northern and central Scotland. Following a rapid decline of more than 60 per cent since the 1970s, the population has now steadied. The reasons for the decline are complicated: the use of modern agricultural methods in the countryside results in there being very little winter stubble and its seed; in towns and cities pollution has reduced the invertebrate prey for the young, and humans have reduced the number of breeding sites due to modern building styles – there are, for example, not enough nesting spaces under eaves.

GREY WAGTAIL ■ *Motacilla cinerea* Length 18–19cm; wingspan 25–27cm

Amber listed

DESCRIPTION This bird constantly wags its tail. In the adult male the upperparts are blue-grey and the rump is yellow-green. There is a white supercilium, a white lower border to the cheeks and a black bib. The underparts are yellow, brightest under the tail, and the wings and tail are black, with a white wing-bar and outer tail. The legs are brownish-pink and not dark as those of other wagtails. The adult female is similar to the male, but her bib is white or mottled black, and her underparts are paler. Juveniles have a white throat, and yellow only under the tail. The similar **Yellow Wagtail**, *Motacilla flava flavissima* a bird of lowland water meadows, is all yellow below and yellowy-green above.

HABITAT Almost confined to swift-flowing, rocky, mostly shallow rivers and streams with wooded banks.

HABITS The diet consists mainly of insects, and Grey Wagtails are often seen flycatching. Nests in a hollow in a riverbank; the female lays 4–6 whitish eggs that are faintly marked with grey. Voice: the call is a distinctive metallic *tzi-tzi*; the song is a pleasing warble, sung in flight or from a perch overhanging a stream.

DISTRIBUTION Because of its rather specialized habitat, this species is most likely to be seen in western and northern Britain; it is very scarce in central and eastern England. In winter it also occurs by lowland lakes and streams, and this is when it is likely to be seen in gardens that are near streams.

adult, winter; summer female similar but grey of back like summer male

PIED WAGTAIL ▪ *Motacilla alba* Length 18cm; wingspan 25–30cm

DESCRIPTION This is a well-named bird, especially in the case of the male, which has a black crown, nape, upperparts, rump. throat and chest, and a white face and principally white underparts. It has white wing-bars, white edges to the tertials and white outer feathers to the black tail. The female has dark grey upperparts and wing-coverts, and a black rump. Both sexes have well-patterned pied wings. The Pied Wagtail is a subspecies of the continental White Wagtail, which has a pale grey back.

HABITAT Occurs especially around farms and cultivation with nearby water.

HABITS The Pied Wagtail is a ground feeder on insects; it walks with a nodding head and wagging tail. It is a territorial breeder, but is communal in autumn and winter, when it is sometimes found roosting in trees in warm city centres, and walking in suburban streets looking for invertebrates in gutters. It nests in hedge banks and crevices in walls. The female lays 5–6 greyish-white eggs that are finely speckled grey-brown. Voice: the call is *tschizzick*; the song consists of lively twittering or warbling.

DISTRIBUTION The species breeds throughout the British Isles, but is less abundant in Scotland except in the east and south. After breeding birds disperse southwards or are short-distance migrants, and are then most common in lowland areas.

male, summer

male, winter

male, White Wagtail

MEADOW PIPIT ■ *Anthus pratensis* Length 15cm; wingspan 22–25cm

Amber listed

DESCRIPTION This species is grey-brown above with an olive tint; it has dark brown stripes on the crown and back. The centres of the coverts and tertials are blackish with pale edges, and the wing-bars are dull. The underparts are white with buff at the sides; the breast and flanks are streaked, and the outer-tail feathers are white.

HABITAT Favours heaths, rough grassland, moors and sand-dunes. It deserts uplands in winter for farmland and the seashore.

HABITS The Meadow Pipit's diet consists mainly of invertebrates, which it finds as it walks steadily to and fro on its yellowish legs and feet. It nests on the ground in a well-concealed, neat nest in heather or long grass. The female lays 3–5 grey to reddish eggs, well-marked blackish. Voice: the call is a weak *tsip tsip*, often uttered in flight. The species sings in flight: initially it utters weak notes, which gather speed as the bird rises and end in a trill as it 'parachutes' down on uplifted wings.

DISTRIBUTION This pipit breeds abundantly in Scotland, Northern Ireland, Wales, northern England and the moors of the West Country. It breeds only patchily in central and southern England. Birds migrate from the Continent to winter in the British Isles. It is most likely to occur in gardens close to its ideal habitat, or flying over on migration, when it may be identified by its distinctive call.

1st autumn

GOLDFINCH ■ *Carduelis carduelis* Length 12cm; wingspan 21–25.5cm

DESCRIPTION In this lovely and unmistakable bird the sexes are similar, with a red, white and black-striped head, the black reaching onto the crown, the white joining under the red chin. The upperparts, breast and flanks are sandy-brown, the rump is white, and the wings and tail are black; the slightly forked tail is white tipped, and the wings have a broad, golden-yellow band right across them and white-tipped flight feathers. The bill is noticeably pointed. The juvenile's greyish-buff head and body look like those of a different species until the autumn moult; the wings and tail are like those of adults.

HABITAT Occurs in orchards, gardens, fringes of woods and commons, wherever there is a good food source of tall weeds, especially Compositae.

HABITS Goldfinches are adept at feeding on seeding thistles, dandelions and teasels. They are gregarious outside the breeding season, often with other finches. In recent years they have become increasingly common at garden feeders, and six or more birds may squabble at a feeder for niger seeds or sunflower hearts. They nest in trees or tall bushes, building a neat cup towards the end of a branch; the female lays 4–6 blue-grey eggs marked with reddish-brown. The birds now even nest in parks and suburbia. Voice: the call is a tri-syllabic *tswitt-witt-witt*, like tiny bells. The song is also tinkling, with trills, and is sung from a prominent song post.

DISTRIBUTION Breeds and winters widely everywhere except north-west Scotland and the Northern Isles, and absent generally from mountains and moorland. After a decline in the 1970s and '80s, the breeding population has increased greatly, especially spreading northwards in Scotland. There has been a more than 90 per cent increase in the population since the 1990s, and up to 60 per cent of gardens in the BTO Garden Birdwatch scheme now record Goldfinches, as the birds have discovered the seeds put out for them. Many winter within the breeding area, but juveniles in particular, and females, go as far as south as the Mediterranean.

juvenile

BULLFINCH ■ *Pyrrhula pyrrhula* Length 14.5–16.5cm; wingspan 22–29cm

Amber listed

DESCRIPTION The male is easily identified: he has a black cap, and an ash-grey mantle and wing-coverts, and is bright pink-red from the ear-coverts to the belly; the under-tail is white, and the wings are black with a white bar, the black tail contrasting with the white rump. The female has the same pattern, but her underparts are grey washed with brown. The juvenile is like the female without the black cap.

HABITAT Found in mixed woods, parks and gardens with good undergrowth or thick hedges.

HABITS The Bullfinch is a quiet, rather secretive species. It is most often seen when disturbed, when it flies off showing its distinctive black tail edged by the white rump. The species' main food is tree buds, especially those of oaks, willows and hawthorns. It is increasingly visiting birdfeeders, but is easily disturbed by other species. It builds a rather shallow nest in a thick, especially evergreen, bush; the female lays 4–6 greenish-blue eggs with dark purple streaks at the large end. Voice: the most common call note is a repeated, subdued *phew* with a falling pitch; good hearing is required to hear the call. The slow, low-pitched song is mostly so quiet that many birdwatchers have never heard it.

DISTRIBUTION The Bullfinch is mostly resident and sedentary, with pairs staying together. It is widespread in summer and winter, but is not found in highlands, the Outer Hebrides or Northern Isles, or open country with few trees. It is most abundant in suitable habitat south of a line from Lancashire to north Yorkshire, and has declined in south-east England.

CHAFFINCH ■ *Fringilla coelebs* Length 14.5cm; wingspan 24.5–28.5cm

DESCRIPTION The male is multi-coloured and very distinctive, and not to be dismissed as 'just another Chaffinch' simply because it is common. It has a blue-grey crown and nape, a black forehead, pink cheeks and underparts, a chestnut mantle, a yellowish-green rump, black wings with an intricate pattern of a white shoulder and wing-bar, yellow-edged secondaries and tertials, and a black tail with white outer feathers. The female has an unmarked buffish-grey head and underparts, less bright wing marks and an olive-brown mantle. In both sexes the white is very distinctive as they fly away.

HABITAT Common in deciduous and coniferous woodland, parks and gardens.

HABITS The Chaffinch is gregarious in winter, when it is often found with other finches, buntings and sparrows. It is regular at garden feeding stations, often numbering in double figures. In winter it is a ground feeder on many kinds of seeds; in summer it takes insects and caterpillars found in trees and bushes. It is a territorial breeder, and constructs a beautiful, neatly made, cup-shaped moss nest lined with hair in a bush, in which the female lays 3–4 pale blue eggs streaked and blotched with purple-brown. Voice: the call is a metallic *chink*; the male's spring call is a monotonous, repeated *wheet*; the tuneful song ends in a flourish. If the song is listened to carefully, it becomes apparent that it repeats a 'tune' several times, then sings a subtly different 'tune'.

DISTRIBUTION This is one of Britain's most common birds, breeding and wintering everywhere from northern Scotland to the Scilly Isles. It is least common on exposed highlands and in the Northern Isles of Scotland. British-bred birds are sedentary, but winter numbers are swollen by visitors from mainland Europe, especially Scandinavia. It is noticeable that in contrast to other garden birds such as the Starling and House Sparrow, the Chaffinch population has increased by over 30 per cent in the last 40 years or more.

BRAMBLING ■ *Fringilla montifringilla* Length 14cm; wingspan 25–26cm

DESCRIPTION The summer male is distinctive: he has a glossy black head, mantle and bill; an orange throat, breast and shoulders; white on the rest of the underparts; two white wing-bars on mostly black wings, and a forked black tail. In flight he shows a noticeable white rump. The female has a greyish head with a buffish supercilium and blackish striped crown; she retains the male's tail and rump pattern, but her wing and body colours are much duller. In a mixed flock with Chaffinches, both sexes of Brambling can be clearly distinguished from their 'cousins' by their orange breast and, when they fly, their long white rump.

HABITAT Breeds in open birch and mixed forests. Winters especially in beech woods, feeding on fallen seed (beechmast), and on stubble, often with other finches.

HABITS Appears in gardens at feeding stations, most often in cold weather, accompanying Chaffinches. Visitors come daily until the weather eases. The common call of migrating birds attracts attention – a rather nasal, wheezy *tsweep*.

DISTRIBUTION The Brambling is a winter visitor to Britain from Scandinavia. Large numbers winter throughout Britain, mostly in England, Wales, and southern and central Scotland. They tend to be less common the further west you are. The species' abundance in any area depends on the size of the autumn influx and the availability of food. If the wild beechmast crop is poor this will increase the chance of the birds wandering and visiting garden feeders.

winter plumage

winter plumage

COMMON CROSSBILL ■ *Loxia curvirostra* Length 16.5cm; wingspan 27–30.5cm

SCOTTISH CROSSBILL ■ *Loxia scotica* Length 15.5cm–17cm; wingspan 27–30cm

Amber listed

DESCRIPTION The Scottish Crossbill is Britain's only endemic species. It can be reliably identified only by careful study of its distinctive call. Both these crossbill species look very alike. They are large, well-built finches with seemingly large heads. The males are very distinctive, with a brick-red head and body, and brown wings; juveniles are often not completely red, and have yellow or orange patches. Females are grey-green with diffuse streaked upperparts, and a yellower rump. Juveniles are like a heavily streaked and browner female. The bill is longer than deep with unique crossed mandibles, the upper one curved more than the lower one.

HABITAT Breeds in conifers, especially Norway Spruce and larch.

HABITS The main food is spruce seeds. Crossbills are acrobatic feeders, using the bill and feet like parrots. A cone's scales are opened with the bill's tips and the seeds are extracted with the tongue, so the fallen cone can be easily distinguished from the chewed remains of a cone left by a rodent. The birds can often be watched closely, drinking from a puddle, garden bird bath or stream. They breed according to the availability of coniferous seeds. The nests are built high up in trees; they are sometimes built early, even when snow is still present, and can be erected in any month if need be. The female lays 3–4 eggs very like those of the Greenfinch (see pp. 148–149). Voice: the call, often made in flight, is harder in the Scottish Crossbill than in the Common Crossbill, an explosive *chip chip*.

Scottish Crossbill

DISTRIBUTION The Scottish species is found only in north-east Scotland, especially in ancient Caledonian forests from Sutherland to Inverness. The Common Crossbill breeds mostly in Scotland, northern and southern England, breckland in East Anglia and Wales, and is scarce in Northern Ireland. It is resident but nomadic, and irruptive in some years, sometimes in huge numbers, when cone crops fail. At such times birds wander and turn up in woods where they are not normally seen.

♀

LINNET ■ *Carduelis cannabina* Length 13.5cm; wingspan 21–25.5cm

Red listed

DESCRIPTION The summer male has a grey head with a crimson forehead and whitish crescents above and below the eyes. He has a crimson breast, buff flanks, white belly, chestnut mantle and wing-coverts, and white edges to the blackish primaries and tail (these show well in flight). The female is brown above streaked darker, with a spotted throat, buff breast and flanks with dark brown streaks, and white belly; the tail and wings have the diagnostic white as in the male. The winter male is similar to the female.

HABITAT Breeds and feeds on heaths and commons with gorse, coastal scrub and young plantations. In the non-breeding season forms flocks on farmland, rough land inland and coasts.

HABITS Large winter flocks feed on stubble, wasteground and fallow land. Linnets breed in loose colonies. The nest is often in bramble or gorse, and in it are laid 4–6 pale blue eggs with brownish spots and streaks. Voice: the flight call is *chichichit*; the fast warbling song is sung from an open perch.

DISTRIBUTION Mostly resident birds, Linnets are widely distributed in summer and winter, but restricted to low ground in Scotland, and similarly in winter in England and Wales. They suffered a big decline of more than 50 per cent in the 1970s and '80s because of the lack of seed and invertebrate food due to the intensification of farming practices. Not a common bird in the garden, the Linnet is most likely to be seen in rural and suburban gardens close to rough ground; it can become a regular visitor in gardens where a good supply of small seed is kept on the ground.

♀

juvenile

LESSER REDPOLL ■ *Carduelis cabaret* Length 11.5cm; wingspan 22.5cm
Red listed

MEALY REDPOLL ■ *Carduelis flammea* Length 12.5; wingspan 21cm

DESCRIPTION These very similar species were listed as related subspecies until recently, and in 2001 were each made full species. Both are brown above, white below, heavily streaked darker on the wings, breast and flanks. The Lesser has a buff tone, and the Mealy is paler and greyer; so the Lesser's wing-bars are buff, while the Mealy's are whitish. Both species have small, pointed bills, tiny black bibs and pinkish-red foreheads; males in summer have red breasts. The two species are hard to tell apart; careful observation is needed.

HABITAT Occurs in willow, birch, alder and juniper forests, and especially in the British Isles, in conifer forests.

HABITS Both species are gregarious outside the breeding season. They often associate with Siskins (see pp. 150–151) in the winter. They are most likely be seen in gardens if they arrive with Siskins, and if their favourite trees are located in the garden or nearby. Here they will search for seeds, especially those of birch and alder, often hanging likes tits do. They often nest high in a tree, and produce 4–6 blue eggs that are streaked with red-brown. Voice: the distinctive call, often made in flight, is a twittering trill, *chuch-uch-uch-uch*.

DISTRIBUTION The Lesser breeds in Scotland, northern England, Northern Ireland and Wales; the northern birds winter further south, mostly in the British Isles. The Mealy is a winter visitor from the boreal zone of northern Europe, Greenland and Iceland. It is subject to irruptions when it is searching for food, and is most likely to be seen then in eastern and central Britain. There are just a handful of breeding records in northern Scotland each year.

adult Lesser

adult Mealy (formerly known as Common Redpoll)

adult Lesser

adult Lesser

GREENFINCH ■ *Carduelis chloris* Length 15cm; wingspan 24.5–27.5cm

DESCRIPTION This is a plump, fork-tailed bird with a stout, conical bill. The male is particularly green – olive-green above, yellowy-green below, with a grey patch on the bases of the secondaries and yellow bases to the primaries; he has a blackish tail with a bright yellow base. The female is duller, browner above, and paler and greyer below. The juvenile is like the female but even less green and heavily streaked above and below.

HABITAT Occurs on the borders of woods, on seed-bearing trees, and is well adapted to urban life in parks, gardens and cemeteries.

HABITS In spring the male has a circular song flight with fluttering wings. The species is gregarious in winter, flocking with other seed eaters. The birds feed on the ground on a great variety of seeds, and are even able to crack open hard hornbeam seeds. The males especially are a fine sight at garden feeders, where they will often drive other birds away. The nest is a bulky cup constructed in a tall bush; the female lays 4–6 off-white eggs sparsely streaked and spotted red-brown and grey. Voice: the male's call is a drawn-out, nasal *tswee*; the flight call is *teu teu*.

DISTRIBUTION The Greenfinch is mostly resident, and widely distributed across Britain, even in the Northern Isles, but not in the uplands of Scotland. In the past 30 years it has increased its range in summer and winter into northern Scotland. It is most abundant all year round on lower ground in central, southern and eastern England. The winter population is increased by small numbers from mainland Europe.

juvenile

SISKIN ■ *Carduelis spinus* Length 12cm; wingspan 20–23cm

DESCRIPTION A tit-sized bird, the male Siskin has a yellow face (with greyish-green ear-coverts), breast, rump and outer tail. The crown and chin are black, and the tail is forked. The wing is an intricate pattern of black, bright yellow and green, with a particularly broad yellow bar across the tips of the coverts and bases of the primaries. The female is patterned like the male but duller, and without the black cap. The juvenile is like a dull female, but is more heavily streaked.

HABITAT Breeds in extensive coniferous and mixed forests, so is most abundant in Scotland, Wales and south-west England; it occurs only in scattered pockets elsewhere in England. In winter it occurs especially in wooded river valleys.

HABITS The Siskin feeds on tree seeds in particular, searching acrobatically in birch, alder and spruce, and feeding less often on the ground than other finches. It is most likely to be seen as a winter visitor in the garden; for a small bird it is very aggressive both to birds of its own species and larger visitors to feeders. It is very fond of peanuts in red bags, niger seeds and sunflower hearts. It nests in a fir or pine, building the nest preferably well out on a horizontal branch; the female lays 4–5 pale blue eggs with red-brown spots. Voice: the call, often made in flight, is a clear *tsüü*; the song is a non-stop twitter ending in a *wheeze*, an attractive sound made from a tree-top if there is a group of birds in the garden.

DISTRIBUTION The species breeds in Scotland, Wales and south-west England, where plantations of its favoured trees are most common; it only breeds in scattered pockets elsewhere in England. Continental birds winter extensively in western Europe, so Siskins are even more widespread in Britain in winter. There has been a great increase in the summer and winter ranges since the 1970s, helped by maturing plantations for breeding, and the ability of winter visitors in the early days (and since) to take advantage of garden birdfeeders.

1st winter

winter male

CIRL BUNTING ■ *Emberiza cirlus* Length 15.5cm; wingspan 22–25.5cm

Red listed

DESCRIPTION The male is similar to the male Yellowhammer (see opposite) on the upperparts and wings, but has a yellowish-grey, lightly streaked rump; most noticeably its head is boldly marked with a black eye-stripe and bib, and it has yellow above and below the eye-stripe, a yellow band below the bib and an olive breast. The rest of the underparts are yellow streaked with chestnut on the flanks. The female is more greyish than the male, or a female Yellowhammer, but has a duller version of the head pattern.

HABITAT Breeds in pastureland with good hedgerows, and grassy downs. Winters on stubble or fallow fields. Now occurs mostly by or near the south-west coast.

HABITS The Cirl Bunting feeds on invertebrates in the summer, and on small seeds during winter, when it tends to feed in flocks. The nest is in dense cover on the ground, and the female lays 2–5 eggs that show the characteristic hair-like markings of buntings. The call is a quiet *sip*. The song is a bold, metallic rattle, usually sung from a prominent perch on top of a bush. It can be confused with the song of the Lesser Whitethroat, but when you are closer to that bird you will notice that its rattle (which can carry for a distance of 200m) is preceded by a low warble, audible from only a few metres away.

DISTRIBUTION Resident in Britain, in the most northerly part of its European range; until the mid-20th century it was found as far north as Yorkshire. There was a great decline in the species in recent years until only a few pairs were left in south Devon. Loss of winter stubble was established as the cause of the decline. Since then targeted conservation by the RSPB and Natural England has raised the population in Devon from about 100 pairs to more than 800, and it has been reintroduced to Cornwall. Farmers, local bird societies, keen garden birdwatchers and schools have all been involved.

YELLOWHAMMER ■ *Emberiza citrinella* Length 16–16.5cm; wingspan 23–29.5cm

Red listed

DESCRIPTION The summer male Yellowhammer is distinctive. He has a yellow head – it is often almost unmarked, but at other times the crown and ear-coverts are greenish-brown. He has warm brown upperparts, streaked darker, a chestnut rump that shows well especially in flight, yellow underparts with chestnut on the sides of the breast, and streaky flanks; the wings and tail are brown-black, the latter with white outer feathers. The female is tricky to tell from other female buntings: she is duller, browner above and less yellow below, and more heavily marked with olive on the crown ear-coverts. Her underparts are more heavily marked, blackish rather than chestnut, and her rump is reddish-brown.

HABITAT Breeds on farmland, heaths, and coastal fields and scrub. Winters on stubble, fallow land and stackyards.

HABITS This species is often seen as a roadside bird, in arable country. It nests in a bank, at the base of a hedge, or in a thick bramble clump or bracken. The female lays 3–5 whitish eggs with a maze of dark scribbles. Voice: the call is a ringing *zit*. The song is commonly rendered as 'little bit of bread and no cheese', a succession of notes ending with a *wheeze*, usually from a prominent song post.

DISTRIBUTION The Yellowhammer occurs almost throughout Britain in suitable habitat. It is closely linked with cereal farming. There has been a decline in numbers of about 50 per cent since the 1970s, probably due to intensification of farming practices and management of grassland, which results in less food, especially invertebrates to feed chicks and, later, seeds in winter. The species mostly winters within the breeding area, and is most likely to occur in rural gardens in central and eastern England, and eastern Scotland, in winter. It is most likely to be encountered in a garden if this borders farmland, where the birds' song can be heard; winter wanderers may come to feeders with finches.

REED BUNTING ■ *Emberiza schoeniclus* Length 15–16.5cm; wingspan 21–28cm

Amber listed

DESCRIPTION The male is very distinctive, with a black head surrounded by a white collar joining a white moustache, and a black bib; the rest of the underparts are whitish with indistinct grey-streaked flanks. The upperparts are mostly brown with black-and-buff streaks, and the rump is grey; the wings are dark brown with chestnut coverts and feather edges. The female is quite difficult to separate from other buntings: she has a brown crown and ear-coverts, creamy supercilium, white moustache and black malar stripe. The rest of her underparts are white streaked with dark brown; the upperparts and wings are as in the male. Juveniles are similar to females.

HABITAT The Reed Bunting is well named because traditionally it occurs in reeds in reedbeds, in rushes, in thick cover by lakesides and riverbanks, among willows by moorland streams and in coastal marshes. More recently it has been occurring on farmland in ditches, in crops like oilseed rape and in young forestry plantations.

HABITS The species is sociable in winter on dry ground, stubble and dunes. Its diet consists mostly of grass and weed seeds; invertebrates are taken in summer. It nests near water, on or near the ground in grass or rushes. The female lays 4–5 olive-brown eggs that are boldly streaked and spotted. Voice: sings from an exposed perch, repeatedly uttering a monotonous (to our ears) *tweek tweek tweek tititik*; the call is a high, soft *seeoo*.

DISTRIBUTION The Reed Bunting is found widely in suitable habitat in summer and winter. It avoids barren uplands, and is most abundant in northern and central England, mostly in lowland areas. Most birds are sedentary in the British Isles. The species has declined overall in southern England, but its winter range is wider than was once believed. It is most likely to occur as a rural garden visitor in winter with finches, and in poor weather even in suburban gardens among wandering finch flocks.

▪ FURTHER INFORMATION ▪

USEFUL ADDRESSES AND FURTHER READING

BirdLife International
Each country has a national birdwatching society which is affiliated as a 'Partner' to BirdLife International, the leading bird research and conservation organization in the world. Their website is www.birdlife.org or write to them for information about the society at their headquarters:
Wellbrook Court
Girton Road
Cambridge
CB3 0NA
UK

RSPB
In the United Kingdom the 'Partner' and leading bird protection organization is:
The Royal Society for the Protection of Birds
The Lodge
Sandy
Bedfordshire
SG19 2DL
www. RSPB.org.uk.

BTO
If you are interested in helping as a volunteer on a bird survey to do with populations, movements or ecology, contact the United Kingdom's expert in that field:
The British Trust for Ornithology
The Nunnery
Thetford
Norfolk
IP24 2PU
www.bto.org

For further reading the following will be particularly useful:

Balmer, D.E., Gillings, S., Caffrey, B.J., Swann, R.L., Downie, I.S., and Fuller, R.J. 2013. *Bird Atlas 2007-2011: The Breeding and Wintering Birds of Britain and Ireland*. BTO Books.
Burton, J. & Young, S. 2005. *The Ultimate Birdfeeder Handbook*. New Holland Publishers.
Cramp. S and Perrins, C M (eds). 1977-1994. *The Birds of the Western Palearctic*, vol. 1-9, Oxford University Press.
du Feu, C. 2003. *The BTO Nestbox Guide*. BTO Books.
Golley, M. and Moss, S. 2011. *The Complete Garden Bird Book*. New Holland Publishers.
Gooders, J. 1994. *Where to Watch Birds in Britain and Europe*. Christopher Helm.
Harrap, S. and Redman, N. 2003. *Where to Watch Birds – Britain*. Christopher Helm.
Holden, P. and Greeves, T. 2002 (4th edition). *RSPB Handbook of British Birds*.
Milne, P. and Hutchinson, C. 2009. *Where to watch Birds – Ireland*. Christopher Helm.

Moss, S. 2011. *Birds Britannia – How the British Fell in Love with Birds*. HarperCollins.
Mullarney, K., Svensson, L. et al. 1999 (2nd edition). *Collins Bird Guide*. HarperCollins. Now available as an App for iphone and ipad
Tucker, G., Heath, M. and Tomialojc, L. 1994, *Birds in Europe, their Conservation Status*. BirdLife International.
Toms, M. 2003. *The BTO/CJ Garden BirdWatch Book*. BTO Books.

There are many books available now which deal in detail with one family or a single species (monographs). Famous ones over the years have been:

Edward A. Armstrong's *The Wren* (1955, Collins).
Paul Donald's *The Skylark* (2004, Poyser).
David Lack's *The Life of the Robin* (1946, Witherby & 1953, Penguin).
J.D.Summers-smith's *The House Sparrow* (1963, Collins).
N.B. Davies' *Dunnock behaviour and Social Evolution* (1992, Oxford University Press).
Derek Ratclffe's *The Peregrine Falcon* (1980, Poyser).

Many county bird societies have published their own detailed book about the county's birds. See your local library for details, or the society's website (Google *Devon Birds Society* or similar, inserting instead your county name).

Other helpful websites are:

www.nestbox.com – sales of nest boxes designed for particular species
www.haiths.com – bird food specialist since 1937
www.birdfood.co.uk – CJ foods, founded 1987, is the largest supplier of bird food in Europe. Supporter of the BTO Garden BirdWatch Survey.
shopping.rspb.org.uk/birds-wildlife/nestboxes
ukwildlifecameras.co.uk

and especially:

www.bto.org/volunteer-surveys/gbw for details of the long-running Garden Birdwatch survey run by the British Trust for Ornithology.

Garden Bird Checklist and Spotter's Guide

Tick boxes are included so that you can mark
off species that you have seen.

Wildfowl (Anatidae)
☐ Canada Goose *Branta canadensis*
☐ Mallard *Anas platyrhynchos*

Game Birds (Phasianidae)
☐ Common Pheasant *Phasianus colchicus*
☐ Grey Partridge *Perdix perdix*
☐ Red-legged Partridge *Alectoris rufa*

Herons (Areidae)
☐ Grey Heron *Ardea cinerea*

Hawks and Allies (Accipitridae)
☐ Common Buzzard *Buteo buteo*
☐ Sparrowhawk *Accipiter nisus*
☐ Red Kite *Milvus milvus*

Falcons (Falconidae)
☐ Kestrel *Falco tinnunculus*
☐ Peregrine Falcon *Falco peregrinus*

Rails (Rallidae)
☐ Moorhen *Gallinula chloropus*

Plovers (Charadriidae)
☐ Lapwing *Vanellus vanellus*

Gulls (Laridae)
☐ Black-headed Gull *Chroicocephalus
ridibundus*
☐ Common Gull *Larus canus*
☐ Lesser Black-backed Bull *Larus fuscus*
☐ Herring Gull *Larus argentatus*
☐ Great Black-backed Gull *Larus marinus*

Pigeons and Doves (Columbidae)
☐ Rock Dove or feral pigeon *Columba livia*
☐ Stock Dove *Columba oenas*
☐ Woodpigeon *Columba palumbus*
☐ Collared Dove *Streptopelia decaocto*

Cuckoos (Cuculidae)
☐ Cuckoo *Cuculus canorus*

Parrots (Psittacidae)
☐ Ring-necked Parakeet *Psittacula krameri*

Barn Owls (Tytonidae)
☐ Barn Owl *Tyto alba*

Typical Owls (Strigidae)
☐ Tawny Owl *Strix aluco*
☐ Little Owl *Athene noctua*

Swifts (Apodidae)
☐ Common Swift *Apus apus*

Woodpeckers (Picidae)
☐ Green Woodpecker *Picus viridis*
☐ Great Spotted Woodpecker *Dendrocopos
major*

Crows (Corvidae)
☐ Jackdaw *Corvus monedula*
☐ Rook *Corvus frugilegus*
☐ Carrion Crow *Corvus corone*
☐ Hooded Crow *Corvus cornix*
☐ Raven *Corvus corax*
☐ Jay *Garrulus glandarius*
☐ Magpie *Pica pica*

Kinglets (Regulidae)
☐ Goldcrest *Regulus regulus*
☐ Firecrest *Regulus ignicapillus*

Typical Tits (Paridae)
☐ Great Tit *Parus major*
☐ Blue Tit *Cyanistes caeruleus*
☐ Coal Tit *Periparus ater*
☐ Marsh Tit *Poecile palustris*
☐ Willow Tit *Poecile montanus*
☐ Crested Tit *Lophophanes cristatus*

Larks (Alaudidae)
☐ Skylark *Alauda arvensis*

Swallows and Martins (Hirundinidae)
☐ Swallow *Hirundo rustica*
☐ House Martin *Delichon urbica*

Long-tailed Tits (Aegithalidae)
☐ Long-tailed Tit *Aegithalos caudatus*

Warblers and Allies (Sylviidae)
☐ Garden Warbler *Sylvia borin*
☐ Blackcap *Sylvia atricapilla*
☐ Grasshopper Warbler *Locustella naevia*
☐ Willow Warbler *Phylloscopus trochilus*
☐ Common Chiffchaff *Phylloscopus collybita*

Waxwings (Bombycillidae)
☐ Waxwing *Bombycilla garrulus*

Nuthatches (Sittidae)
☐ Nuthatch *Sitta europaea*

Treecreepers (Certhidae)
☐ Tree creeper *Certhia familiaris*

Wrens (Troglodytidae)
☐ Wren *Troglodytes troglodytes*

Starlings (Sturnidae)
☐ Starling *Sturnus vulgaris*

Thrushes (Turdidae)
☐ Song Thrush *Turdus philomelos*
☐ Mistle Thrush *Turdus viscivorus*
☐ Blackbird *Turdus merula*
☐ Fieldfare *Turdus pilaris*
☐ Redwing *Turdus iliacus*

Chats (Turdidae)
☐ Robin *Erithacus rubecula*
☐ Nightingale *Luscinia megarhynchos*
☐ Common Redstart *Phoenicurus phoenicurus*
☐ Black Redstart *Phoenicurus ochruros*
☐ Stonechat *Saxicola torquata*

Flycatchers (Muscicapidae)
☐ Spotted Flycatcher *Muscicapa striata*

Accentors (Prunellidae)
☐ Dunnock *Prunella modularis*

Sparrows (Passeridae)
☐ House Sparrow *Passer domesticus*
☐ Tree Sparrow *Passer montanus*

Wagtails and Pipits (Motacillidae)
☐ Grey Wagtail *Motacilla cinerea*
☐ Pied Wagtail *Motacilla alba*
☐ Meadow Pipit *Anthus pratensis*

Finches (Fringillidae)
☐ Bullfinch *Pyrrhula pyrrhula*
☐ Chaffinch *Fringilla coelebs*
☐ Brambling *Fringilla montifringilla*
☐ Common Crossbill *Loxia curvirostra*
☐ Scottish Crossbill *Loxia scotica*
☐ Linnet *Carduelis cannabina*
☐ Lesser Redpoll *Carduelis cabaret*
☐ Mealy Redpoll *Carduelis flammea*
☐ Goldfinch *Carduelis carduelis*
☐ Greenfinch *Carduelis chloris*
☐ Siskin *Carduelis spinus*

Buntings (Emberizidae)
☐ Cirl Bunting *Emberiza cirlus*
☐ Yellowhammer *Emberiza citrinella*
☐ Reed Bunting *Emberiza schoeniclus*

■ INDEX ■

160